Electronics for the Electrician

By Newton C. Braga

Other Prompt® books
by Newton C. Braga

Sourcebook for Electronics Calculations, Formulas, and Tables

Fun Projects for the Experimenter

Electronics for the Electrician

By Newton C. Braga

Electronics for the Electrician

©2000 by Sams Technical Publishing

PROMPT© Publications is an imprint of Sams Technical Publishing, 5436 W. 78th St., Indianapolis, IN 46268.

All rights reserved. No part of this book shall be reproduced, stored in a retrieval system, or transmitted by any means, electronic, mechanical, photocopying, recording, or otherwise, without written permission from the publisher. No patent liability is assumed with respect to the use of the information contained herein. While every precaution has been taken in the preparation of this book, the author, the publisher or seller assumes no responsibility for errors or omissions. Neither is any liability assumed for damages resulting from the use of information contained herein.

International Standard Book Number: 0-7906-1218-6
Library of Congress Catalog Card Number: 00-108430

Acquisitions Editor: Alice J. Tripp
Editor: Cricket Franklin
Assistant Editor: Kim Heusel
Typesetting: Kim Heusel
Indexing: Kim Heusel
Cover Design: Christy Pierce
Graphics Conversion: Bill Skinner, Phil Velikan, Christy Pierce
Illustrations: Courtesy the author

Trademark Acknowledgments:
All product illustrations, product names and logos are trademarks of their respective manufacturers. All terms in this book that are known or suspected to be trademarks or services have been appropriately capitalized. PROMPT® Publications and Sams Technical Publishing cannot attest to the accuracy of this information. Use of an illustration, term or logo in this book should not be regarded as affecting the validity of any trademark or service mark.

PRINTED IN THE UNITED STATES OF AMERICA

9 8 7 6 5 4 3 2 1

Electronics for the Electrician

By Newton C. Braga

Electronics for the Electrician

©2000 by Sams Technical Publishing

PROMPT© Publications is an imprint of Sams Technical Publishing, 5436 W. 78th St., Indianapolis, IN 46268.

All rights reserved. No part of this book shall be reproduced, stored in a retrieval system, or transmitted by any means, electronic, mechanical, photocopying, recording, or otherwise, without written permission from the publisher. No patent liability is assumed with respect to the use of the information contained herein. While every precaution has been taken in the preparation of this book, the author, the publisher or seller assumes no responsibility for errors or omissions. Neither is any liability assumed for damages resulting from the use of information contained herein.

International Standard Book Number: 0-7906-1218-6
Library of Congress Catalog Card Number: 00-108430

Acquisitions Editor: Alice J. Tripp
Editor: Cricket Franklin
Assistant Editor: Kim Heusel
Typesetting: Kim Heusel
Indexing: Kim Heusel
Cover Design: Christy Pierce
Graphics Conversion: Bill Skinner, Phil Velikan, Christy Pierce
Illustrations: Courtesy the author

Trademark Acknowledgments:
All product illustrations, product names and logos are trademarks of their respective manufacturers. All terms in this book that are known or suspected to be trademarks or services have been appropriately capitalized. PROMPT® Publications and Sams Technical Publishing cannot attest to the accuracy of this information. Use of an illustration, term or logo in this book should not be regarded as affecting the validity of any trademark or service mark.

PRINTED IN THE UNITED STATES OF AMERICA

9 8 7 6 5 4 3 2 1

Contents

Preface vii
About the Author ix

Section 1

The Differences Between Electricity and Electronics 1
    Fundamentals 2

Section 2

Electronic Components 41
    Wires 41
    Fuses 43
    Switches 45
    Cells and Batteries 47
    Resistors 50
    Variable Resistors 55
    Incandescent Lamps 58
    Neon Lamps 60
    Light-Dependent Resistor or CdS Cells 63

NTC/PTC 65
VDR 67
Capacitors 69
Variable Capacitors 74
Coils or Inductors 76
Transformers 78
Relays 81
Solenoids 84
Motors 86
Loudspeakers and Headphones 88
Magnetic Transducers 91
Piezoeletric Transducers 92
Semiconductors 95
Diodes 97
Zener Diodes 101
LEDs 104
Special Diodes 109
Bipolar Transistors 111
Phototransistors 122
Optocouplers 126
Darlington Transistors 127
Unijunction Transistor (UJT) 132
Programmable Unijunction Transistor—PUT 137
Field-Effect Transistor—FET 140
MOSFETS 145
Power FETs 148
Isolated-Gate Bipolar Transistor—IGBT 152
Silicon Controlled Rectifier—SCR 154
Triac 162
Electromagnetic Interference (EMI) 167
Diac 171
Quadrac 172
Silicon Unilateral Switch—SUS 172
Silicon Bilateral Switches—SBS 174
Liquid Crystal Displays—LCD 175
Tubes 179
Integrated Circuits 184

Section 3

Troubleshooting and Repair 217
    The First Step 218
    Safety Rules of Troubleshooting 218
    The Multimeter 221
    How to Use a Multimeter 223
    Reading and Interpreting Multimeter Values 227
    Finding Parts 227
    Practical Circuits: How They Work 230
    Flashers 237
    Automatic Lighting/Emergency Lighting 243
    Alarms 245
    Doorbells And Chimes 249
    Battery Chargers 250
    Fluorescent Lamps 253
    AC/AC Converters 254
    Power Conditioners 255
    Intercoms 256
    Wireless Systems 258
    High-Frequency Appliances 262
    Surge Protectors 264
    Conclusion 265

Index 267

Preface

The time of the electrician dealing only with simple components such as wires, outlets, switches, and incandescent lamps in his daily work is gone. Each day, electronics extends its branches to other activities. The home and industrial electrician, the automobile electrician, the construction engineer, and the architect are now finding strange components they may not be able to identify, which can cause embarrassing situations.

Electronic components in electric circuits and installations are becoming more common. Timers, dimmers, heater controls, intercom phones, alarms, doorbells, battery chargers, emergency lights, illumination controls, air conditioners, washing machines, and electronic injection systems are all electric devices and appliances where electronic circuits are found. The electrician and other professionals who come into contact with electric installations and who want to be updated with new technologies in these fields must know how these circuits work. This means learning a new science: electronics.

Until some years ago electronics and electricity were independent fields. The electronics technician fixed and repaired only electronic devices, concentrating his efforts in radio and TV, while the electrician worked with cars, home and industrial installations, and energy distribution. They were entirely different fields. Now the electrician must know how a timer or a dimmer works because he needs to know how to repair one if necessary, and a timer is a device of the other field—electronics. Other professionals such as architects and construction engineers must also know about many electronics devices because they must include them in their projects.

Of course, an electronics person, electrician, and other professionals in these areas must know these devices are not as advanced as the ones needed by a telecommunications, hardware, or other advanced specialist in the electronics field. Electricians must know the basic electronics devices found in their work. They must know how to recognize components and devices, how to work with simple circuits, and even detect problems in them. The electrician must know the basic electronics of common circuits used in electric installations of homes, industries, and cars to be secure when installing devices using this technology. The electrician also must know how to examine electronic circuits found in electric installations using basic instruments such as the multimeter.

Where can the electrician and other professionals find basic information about electronics used in electric installations? It was in thinking about this matter that I decided to prepare this book. It gives the reader the fundamentals of electronics, but with an approach familiar to the electrician and professionals who have had some contact with electric installations. The book explains how the basic electronic components work, how they are used in practical applications (homes, cars and industries), and how the electrician can work with them.

This isn't a simple book of electronics directed to all people. It is a book of electronics for electricians and other professionals from construction engineers to architects. It starts with knowledge about electricity and installations that electricians use. Eliminating some steps and going directly to the level where the reader is makes it easier to add important information about electronic circuits and devices. This makes the electrician much more prepared to make money when finding devices with new technology in his work.

This book is intended to give professionals basic knowledge about electronics in order to understand how electronic circuits and devices work in electric installations. The only thing I recommend to the reader is to proceed, and keep in mind that electronics will be increasingly present each day and that electric installations are not the exception.

About the Author

Newton C. Braga has written more than 50 books about electronics and electricity that have been published in his native Brazil as well as Europe and the United States. The main training network of the Brazilian Federation of Industries (SENAI) recommended *Electric Installations For Beginners* and *The Basic Electronics Course,* published in Brazil, as basic texts in its courses.

Electronics for the Electrician

Section I

The Differences Between Electricity and Electronics

The aim of this book is to explain how electronic circuits and devices work without assuming the reader has prior knowledge. Many electricians or professionals of nonelectronic areas probably have some elementary knowledge of electricity. We must assume there are many differences between the work of an electrician and an electronics technician. Not only are the components different, but in many cases the way they work is different as well.

So, it is normal that the first thing the electrician should consider when starting with these studies in electronics is if the knowledge he has about electricity is the same knowledge necessary to learn electronics.

The answer is yes.

The small differences that exist are in the way electricity is used in a domestic, industrial, or automobile installation and the way it is used in an electronic circuit. In essence the electricity is the same.

This means that the basic principles of electricity one needs to know when working on electronic devices and circuits are also the same. The electricians who want to upgrade their knowledge starting with a new science don't have to begin at zero and forget everything studied prior or learned during years of work experience. But, to be successful when learning electronics, it is a good idea to change some points of view about the way the principles are studied, especially when applied to electronics.

Other than electricians who are more practical than theoretical, two other kinds of electricians had to be considered. First, I considered those with a solid base of the theory and apply it in their profession, and secondly, those with a theoretical knowledge that perhaps has been dulled by time. For the first and third groups, a section on the fundamentals of electricity is included.

A section on approaches of the electricity principles applied to electronics is also included for that second group with the solid concept base. Perhaps you may not need this review of basic concepts, but a review is not a bad idea. After all, nobody is free from small slips in memory.

Through a different approach that involves redirecting the ideas, laws, principles, and concepts of electronics, the reader who is familiar with applications in electric installations can form a new image of electricity. This is very important when trying to understand how devices and circuits of a newer science work.

Finally, please realize a complete course of all basic principles needed to know about electronics needs much more space than is available here. Of course, my efforts were directed to include all the essential information a reader needs to have to understand how the main electronic circuits and devices work. If the reader wants to go a step further, much more detailed explanations can be found in many of the books offered by PROMPT®.

Fundamentals

Both electricians and electronics technicians work with the same kind of energy or "fluid" electricity.

As previously stated, the basic differences between an electronic device and an electric device are in the way they are mounted, the parts they

Section I—The Differences Between Electricity and Electronics

use, and the way the electricity is used. Although the ends and the components used to find them are different, the principles of operation of each part of an electronic device and an electric device are the same.

Electricity is not found in different types. This means that the laws of physics that determine how electricity works and can be used are, in essence, universal. This is very important for the electrician who wants to learn about electronics. There are no differences between the laws and principles applied to electronic circuits and devices and the electric circuits and devices.

The Electric Current

Every material substance is made up of small particles called atoms. Atoms are so small that they are not visible under normal conditions. Then scientists discovered that atoms are made of even tinier particles called elementary or atomic particles as shown in Figure 1.

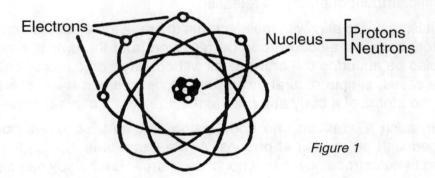

Figure 1

The nucleus of an atom is formed by protons and neutrons surrounded by electrons in an orbital shell. Electrons and protons manifest a special property; they are electrically charged. Electric properties of electrons are negative (−) whereas protons are positively charged (+). Bodies with different charges attract each other, and bodies having charges of the same type, either positive or negative, repel each other. This is why the electrons stay in an orbital movement attracted by the nucleus' charge.

In normal conditions, the number of electrons of an atom is equal to the number of protons. This means the electric charge of the protons and electrons is the same, so the atom is balanced (Figure 2).

Electronics for the Electrician

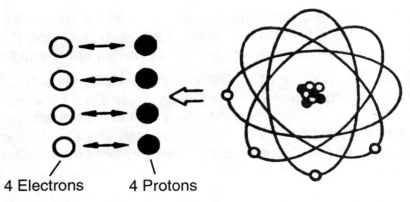

4 Electrons   4 Protons

Figure 2

Although the forces act to maintain an atom in neutral or balanced condition, this stability can be upset under certain conditions. Of course, it is impossible to remove protons from the nucleus of an atom without destroying or changing its nature. This involves special conditions such as those created in a nuclear power plant or by an atomic bomb, where a fantastic amount of energy is released.

It is easier to play with the electrons to break the electrical balance of an atom because electrons can be removed from the orbital shell. They can also be added to the orbital shell without large amounts of energy. In some cases, simple natural or artificial processes can remove electrons from the atoms of a body and put them in the atoms of other bodies.

In such a situation, the body losing electrons becomes positively charged with an excess of protons. At the same time, the body that received the electrons now has a negative charge. Each body has a charge so, we say that the bodies have electric charges or are "electrified." An example of this is shown in Figure 3.

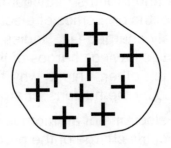 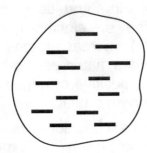

Figure 3

Positive charged body   Negative charged body

Section I—The Differences Between Electricity and Electronics

When you rub a plastic pen or glass rod in a piece of tissue (silk or wool works best), electrons are removed from the atoms of the plastic or glass and transferred to the tissue. Both objects become charged, but with opposite charges as shown in Figure 4.

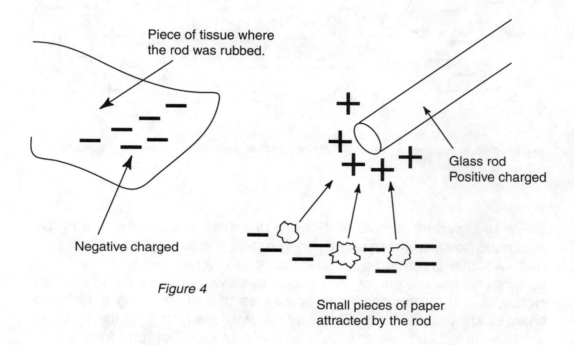

Figure 4

Putting the pen near your hair or small pieces of paper will attract the hair or paper to the pen that is now positively charged. The pen needs the electrons in your hair or the paper to balance its charge. The pen and the piece of tissue have been electrified by attrition, but there are other ways to remove or add electrons to a body, thereby charging them with electricity. The electricity or electric charge in these bodies is fixed or static. The branch of physics that studies the properties of electricity accumulated in objects is static electricity.

Now, going a step farther, if we bridge two charged bodies using a metal wire (a material through which the electrons can move easily—a conductor), a flux of electrons can be noted; the excess electrons in one body move to the body that needs them. This flux of electrons or electron flow is called electric current and is represented in Figure 5.

Electronics for the Electrician

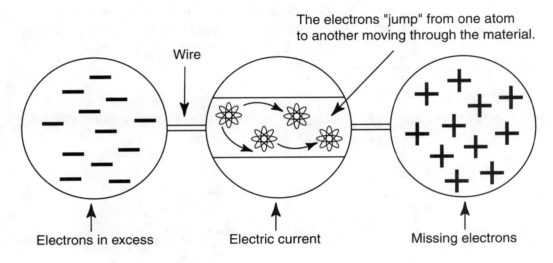

Figure 5

It is important to note that in this process the electrons flowing between the bodies release energy. In this case, the energy is converted into heat. Another important point to observe in this process is that a flux of current can release energy to be used by devices placed along the way. Notice that in all the described processes the only moving particles or charges are the electrons. This explains why the names of the sciences are "electricity" and "electronics" and not "protoncity" or "protonics"!

Scientists found that the existence of atoms with positively charged electrons and negatively charged protons is possible somewhere in the universe. But, these atoms result in antimatter bodies that explode when in contact with our normal atoms. These "antielectrons," or positively charged electrons, are called "positrons."

The current flow between two charged bodies, as shown in Figure 5, is very short. As soon the electrons from one body reach the other body neutralizing any charge, the electron flow stops and the current ceases. The practical use of the energy that is transported by an electric current comes from the fact that we can create ways to maintain the difference in charges between two bodies and maintain a flux of electrons or current between them. Current flow is a dynamic process countering the static electricity.

Section I—The Differences Between Electricity and Electronics

a) Basic Circuits

The electric current can transport energy from one point to another, maintaining the difference of a charge's concentration between the ends of the wire with a current flowing through it.

In the example, two bodies are shown, one with negative charges and the other with positive charges. This is not the only possible configuration that can produce a current flowing between two bodies. They can both be negative or positive but in different degrees or with a different concentration of charges as shown in Figure 6. The charges in these bodies are in different degrees of compression meaning that different repulsion forces move them, either attracting electrons or repelling them.

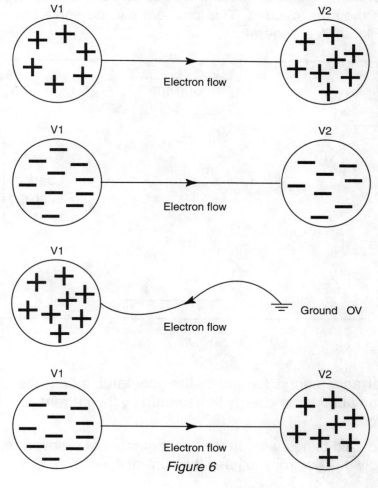

Figure 6

Electronics for the Electrician

If we place a metal wire between these bodies, a current flow will be formed. The current tends to flow from the more compressed negative charges to the less compressed negative charges. This compression force of charges is very important to establish an electric current. In the case of positively charged bodies, electrons will flow from the body with the least amount of positive charge to where the most electronics are missing. As a practical rule, it is easy to see the natural tendency for a balance to be reached.

Returning to our problem, how do we maintain the current if we want to create a constant supply of energy? First, we need an efficient way to replace the electrons that leave the negatively charged body. Since electrons can't be created from nothing, we need a body that can furnish them as long as they are needed. This problem can be solved by the device shown in Figure 7, a generator.

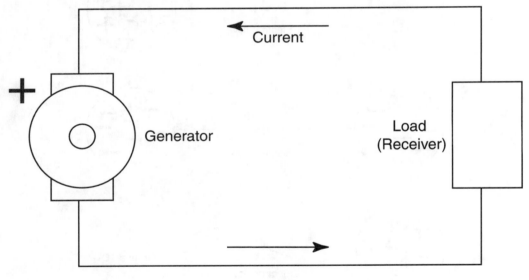

Figure 7

This arrangement is formed by the generator, metal wires, and a device that can absorb the energy transported by the current flow. This is an electric circuit.

So, how do things work in this arrangement and how does the generator supply the electrons to keep the current flowing? The generator can

Section I—The Differences Between Electricity and Electronics

be seen as a two-pole device where one pole is a body with excess electrons (negative pole) and the other has missing electrons (positive pole).

The excess electrons are "pumped" in one direction to the pole where they are missing, passing through a device where the released energy is transformed into heat. Arriving at the pole where the electrons are missing, a replacing process occurs: inside the generator the electrons are taken from the positive pole and replaced to the negative pole again. In this process the generator uses energy. Chemical generators like cells and batteries use energy released in chemical reactions. Mechanical generators known as dynamos use the force applied to a mechanism.

It is important to observe that the generator only needs to replace energy when the current is flowing because in a cell there is no chemical reaction until the moment when it is used. Another important fact to observe is that energy can't be created from nothing and when the current is carrying energy to any device, it must be produced from another kind of energy. The generator doesn't create energy, but transforms one kind of energy into another as suggested in Figure 8.

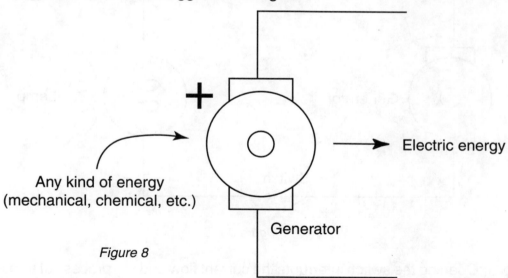

Figure 8

Another important thing to observe in the example is the movement of the electrons and the electric current flowing through a closed loop. This means that the current flow needs a closed circuit to be established. This configuration is called an electric circuit. Going a little bit further with our explanations, a switch can be included in this simple electric circuit to control the current. This is shown in Figure 9.

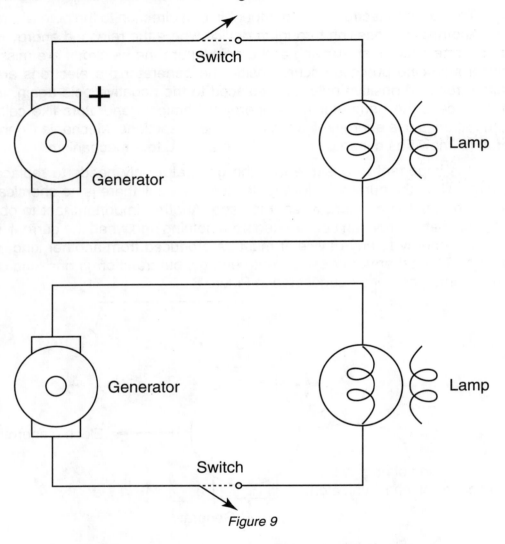

Figure 9

Opening the switch interrupts the current flow and the process of transferring energy to a device. If the device is a lamp we can turn it on and off. Closing the circuit, the current can flow and energy is transferred from the generator to the device; the lamp glows. Opening the switch, no current flows and no energy is transferred to the lamp; it remains off. The important thing to observe in this circuit is the position of the switch. Whether we

Section I—The Differences Between Electricity and Electronics

place the switch after or before the device (having the current flow as reference), the effects are the same. The current can be interrupted in the same manner and the device doesn't receive any energy.

Another point that causes some doubts in electricity students is that the electrons aren't the energy and so they are not consumed when passing through a device. The electrons after and before the device are the same, and the amount of current in the entire circuit is also the same. The electrons are used only to transport energy. They are not the energy!

b) *Voltage and Current*

The excess electrons in a body and the electrons missing in another body represent a state of compression or tension. The situation can be compared to the situation found in compressed springs. The force that tends to recover the balanced state of a body depends on how the charges are compressed as shown in Figure 10.

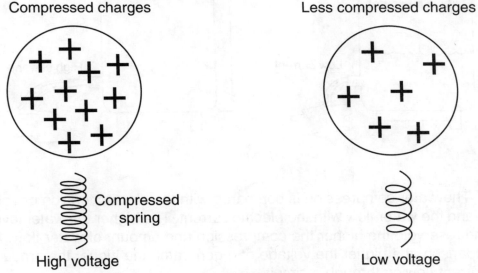

Figure 10

It is this tension that pumps the charges through a metal wire when that body is connected to another. Of course, if the charges in the other bodies are in the same state of tension, the forces pumping the charges are equal and no current flows. It is easy to conclude from this explanation that it is necessary to have a state of tension to produce current. The electric tension is the cause and the current is the effect.

This state of tension or compression is called electric tension or electric voltage and is measured in volts (V).

In a generator, like a cell for instance, there exists a state of tension between its poles—an electric tension. This means that when any circuit is wired between its poles a current can flow. This state of tension in a generator is called electromotive force (emf) and it is also measured in volts. For example, the emf of a single AA cell is 1.5 V. Current is measured in amperes (A).

Figure 11 depicts an analogy between electric current, electric voltage, and the water flowing from a reservoir.

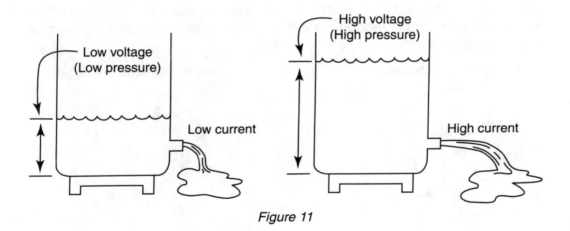

Figure 11

The water compression is compared with the electric tension or voltage and the water flow with the electric current. The higher the water level in the reservoir, the higher the compression and amount of water flow. In comparison, the higher the voltage of a generator, the higher the amount of current flowing through a circuit.

c) Resistance

When describing the basic circuit and the analogy between electric current and water flow, we added an important element: the device that uses the energy released by the generator converting it into heat or light. When reaching this device, the electron flow finds a certain degree of opposition to its movement as suggested in Figure 12.

Section I—The Differences Between Electricity and Electronics

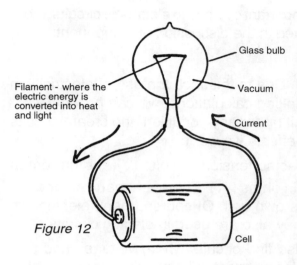

Figure 12

Energy must be released to break this resistance. This energy is converted into heat. The opposition to the current flow is called "electric resistance" and is measured in ohms (Ω). The amount of resistance presented by a device determines how much current can flow through it when a determined voltage is applied to the circuit. The relationship among current, resistance, and voltage in a circuit can be calculated by Ohm's Law (Figure 13).

In a simple manner, considering that the voltage is the cause of a current, we can say that the amount of current flowing through a circuit with a certain resistance is proportional to the applied voltage. The electrician will find that all electronic circuits and devices presenting a resistance will follow this law the same as any other electric device does. In electronic circuits the electrician will also find a device that is used to add a certain resistance to a circuit. It is used in applications where the current flow can be reduced to a determined value. It is a resistor and uses the symbol shown in Figure 13.

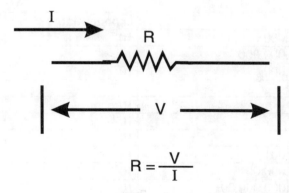

R = Resistance in ohms

I = Current in amperes (A)

V = Voltage in volts (V)

Figure 13

Because resistors are very important devices in electronic circuits, more information about them is presented in the discussion of components.

d) Units

Electricity can be measured. Due to this fact, the use of electricity in many applications is possible. Making calculations, we can preview what happens with electricity in a circuit under any condition and create circuits and devices that produce all the effects we want.

We measure voltages or electric tension in volts (V), the current in amperes (A), and the resistance in ohms (Ω). There are many other electric quantities, each one using its own unit. Of course, many electricians are familiar with these units as they also are used in electric circuits.

Because electricians can also find circuits where voltages and currents are very high or very low, it is important to know how to express high values of current and voltages. Commonly used are the Greek prefixes kilo (k) and mega (M). So, instead of saying 1000 volts, the electrician can say 1 kilovolt (kV) where the kilo (k) represents x 1000. Thus, 1000 volts is also 1 kV.

In electronics high voltages and high resistances are common, but much more common are the small values of currents and other quantities such as capacitances, inductances, etc. To represent the small quantities we can also use Greek prefixes.

Table 1-1 shows the most common Greek prefixes for representing electric quantities:

Prefix	Abbreviation	Times	exponential
Tera	T	1,000,000,000,000	10^{12}
Giga	G	1,000,000,000	10^{9}
Mega	M	1,000,000	10^{6}
Kilo	K	1,000	10^{3}
Deca	D	10	10
Deci	D	0.1	10^{-1}
Mili	M	0.001	10^{-3}
Micro	μ	0.000,001	10^{-6}
Nano	N	0.000,000,001	10^{-9}
Pico	P	0.000,000,000,001	10^{-12}

Table 1-1

In electronic circuits, it is common to see specifications of current like 10μA, meaning 0.000001 A, or specifications of resistances like 2.2 M$_\Omega$, meaning 2,200,000 ohms. In some cases, the decimal point can be replaced by the prefix. So, instead of 2.2 k$_\Omega$ to express 2,200 ohms, we can write 2k2.

e) Effects of Electric Current

The device used as load in our example presents a pure electric resistance or an "Ohmic" resistance, converting all the electric energy into heat. Many devices and electric appliances use this effect of the electric current for heaters, incandescent lamps, etc. But, this is not the only effect of an electric current. When the electrons flow through certain materials or devices other effects can be noted.

Since the electric and electronic devices as well as appliances use all the effects of an electric current it is very important to any electrician to remember what these effects are and how they are used.

Thermal Effect or Joule Effect

When passing through any device or circuit presenting an electric resistance, the result of a current flow is that the energy spent in the process is converted into heat. This effect is also known as Joule's Effect. The amount of energy released can by calculated by Joule's Law.

Joule's Law says that the amount of heat dissipated each second or converted into heat by a device with a constant resistance is given by the voltage applied, multiplied by the current flowing through it.

The energy is measured in joules (J), and the amount of energy in each second is another quantity measured in watts (W). One watt is one joule per second.

P = V x I

Where:

P is the power converted in heat in watts (W)

V is the applied voltage in volts (V)

I is the current through the circuit in amperes (A)

Since current and voltage depend on the resistance, the next formulas are useful to calculate the converted power:

$P = R \times I^2$

and

$P = V^2/R$

Where:

R is the resistance in ohms (Ω)

Be aware there are many other formulas involved in electronic calculations that depend on the application and the complexity of the involved circuit. These are very simple and important to the electrician who wants to start with electronics. To those who want to go on to higher levels and have a useful handbook of electronic formulas and calculations for reference, the *Sourcebook for Electronics Calculations: Formulas and Tables* is an excellent source. (Also written by Newton C. Braga and published by Prompt® Publications.)

Many electronic devices use the Joule Effect as a basis of operation; others have the production of heat only as a result of their operation. Some devices where heat is the basis of operation are incandescent lamps, heaters, and boilers. Other devices produce heat using this effect, but as a consequence of their operation. In this case, the heat must be eliminated in order to avoid causing damage to the device.

Some electronic components producing large amounts of heat must be mounted in heat sinks (Figure 14).

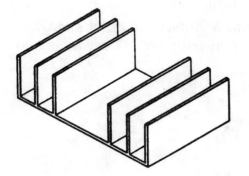
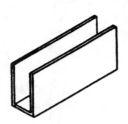

Figure 14

The function of a heat sink is to transfer heat generated by the devices to the air, and in some cases, with the aid of some forced process. Small fans can be used. The electrician will also observe that many circuits are very sensitive to the excess heat produced by components. Heat is the origin of many device failures.

Luminous Effect

There are many ways to produce light from electric energy. The most common way can be considered a consequence of Joule's Effect. In a common incandescent lamp a metal filament (tungsten) is heated by the electric current to a temperature high enough to produce light. Light can also be produced by a current flowing through a gas tube as in neon, xenon and fluorescent lamps. The gas becomes a conductor by ionization and in this process electromagnetic waves in the range of visible, infrared, and ultraviolet are produced. Another way to produce light from electric current is found in electronic components called LEDs (Light Emitting Diodes) whose operation principle is addressed later in the book.

Magnetic Effect

This is the only effect that ever occurs independent of the material in which the electric current flows. A Danish researcher named Oesterd discovered, with a simple experiment, that the movement of electric charges through a wire (current) produces a magnetic field as shown in Figure 15.

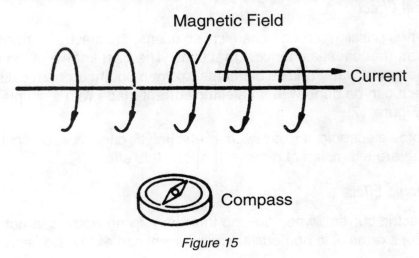

Figure 15

Magnetic field lines involve having the wire oriented in a manner that depends on the direction of the current flow. This effect can be used in many electronic devices. If a current flows through a wire formed into a coil, for example, the magnetic field can be concentrated inside it as shown in Figure 16.

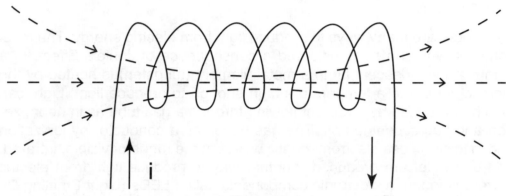

Figure 16

Other researchers found that the inverse effect also exists; if a wire is cut by a magnetic field, an electric current is induced. Devices such as solenoids, electromagnets, relays, motors, loudspeakers, and many others operate based on this effect. These devices include transducers and sensors.

Chemical Effect

When passing through chemical solutions, the electric current can be the agent that causes chemical reactions. The best known of them is decomposition of water and its elements (oxygen and hydrogen) called electrolysis. It can be done with an electric current made with a simple experiment (Figure 17).

Some electronic devices such as electrolytic capacitors and cells when charging are examples of devices that use this effect.

Physiologic Effect

Electric current, when flowing through a living body, can act on that body. For example, in humans electric current can excite the nervous sys-

Section I—The Differences Between Electricity and Electronics

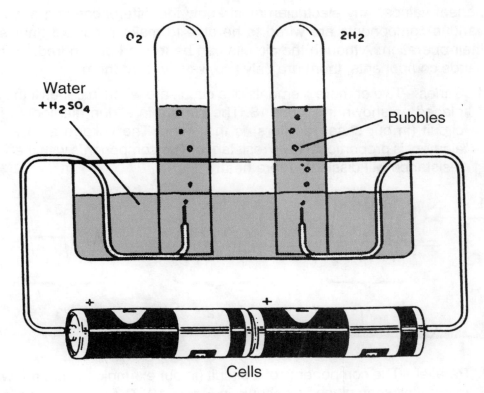

Figure 17

tem. It can cause a little itching sensation when the current flow is very low or severe shocks and burns if the current flow is high. This effect is used in some applications for nerve stimulators and other medical applications.

f) Series and Parallel

In an electric circuit one power supply can be used to power many elements. Even in a simple circuit several elements are used. The manner in which these elements are wired determines how the current flows, how the voltages are divided, and how the heat is produced by each element. In electrical applications the analysis of the circuits is generally simple when few elements are used.

Otherwise, in electronics the circuits used to be very complex, with hundreds or even thousands components. Knowing the way in which the components are wired is fundamental to find failures or to calculate component values. Any electrician must know the differences in the way the various components are wired to be able to make a correct analysis of their operation. Although the circuits can be formed of hundreds or thousands components, there are only two ways to wire them:

1. Series—Two or more elements of a circuit are wired in series if they are placed as shown in Figure 18. The current in all the elements of this circuit (in our case, resistors) is the same. The voltage across each element is proportional to its resistance. The components with the larger resistance will dissipate more heat.

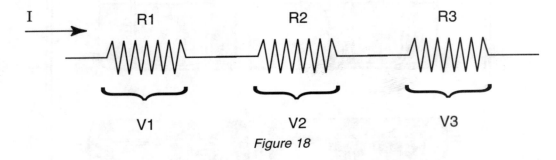

Figure 18

2. Parallel—The components of a circuit (in our example lamps) are wired in parallel when placed as shown in Figure 19. The voltages applied to all the components are the same, and the element presenting the lower resistance dissipates more heat as the current in this element is also the higher.

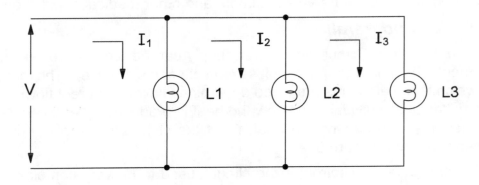

Figure 19

Section I—The Differences Between Electricity and Electronics

When working with electronic circuits, the electrician can find many types of components and, they can be combined in complex associations where some devices are in parallel and others are in series. So, series-parallel associations are common (Figure 20).

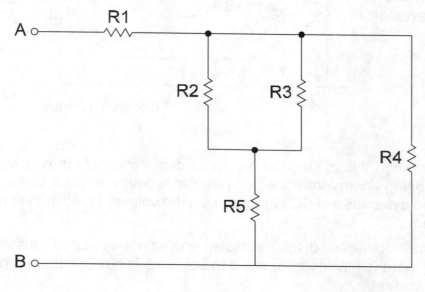

Figure 20

The electrician can analyze one of these configurations taking into consideration which components are in series and which are in parallel. For readers who want to learn more about this topic, a specific electronics course in calculus of equivalence resistance associations is recommended.

g) AC and DC

In basic examples used to show how an electric circuit works, we supposed energy sources or generators with special characteristics. In the poles of these generators we found a constant voltage meaning that when a constant resistance circuit was wired to them, the current flow was established in one direction with a constant value.

This circuit operates with a DC or direct current generator and the current through all its elements is referred to as DC (Figure 21).

Electronics for the Electrician

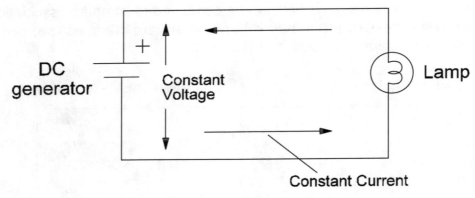

Figure 21

Through this circuit, the electrons can move only in one direction, transporting energy from the generator to a device or load. Cells, batteries, and dynamos are DC generators as the voltage found in their poles is constant.

On the other hand, the electrician who works with electric installations knows that the electricity supplied to buildings is not DC, but AC or alternating current.

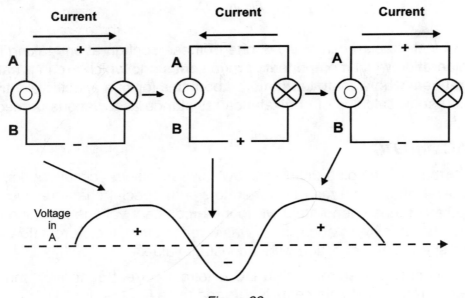

Figure 22

Section I—The Differences Between Electricity and Electronics

What is the difference and what does it mean? It is not necessary to have a constant flux of charges or constant current flow through a device to transport energy to it. Figure 22 shows a generator that changes the voltage in its terminals at high speed.

One moment pole A is positive and B is negative. The electrons are pumped through the device being used as load (a lamp, for example) releasing their energy and heating the filament. The lamp glows. An instant later, the poles of A and B are inverted. The electrons are now pumped in the opposite direction but must pass through the filament of the lamp again releasing energy. The lamp continues to glow. If the generator continues to change its poles, the electrons are pumped forward and backward through the circuit.

Observe that in this process, the charges move only a microscopic distance in one direction, then reverse motion so the charge retraces its path. The action is repeated continually so that the actual average movement of the charges is zero. But, a current flows because the charges are in motion. This is enough to cause an oscillating movement of the electrons in the circuit and to transfer energy from the generator to the load (any device that can absorb energy). This kind of current moving forward and backward is alternating current (AC).

The curious fact to be observed is, although the electrons move only a few centimeters each second, the force that pumps the electrons acts as a wave that moves near the speed of the light through the wires. In domestic installations the AC energy found in outlet jacks changes its poles 120 times by second, or 60 times in a second a pole is positive and 60 times is negative. We say that the frequency of the AC power line is 60 hertz (60 Hz). This is standard in most countries, but in some it is 50 Hz.

The alternating current found in the AC power line has a special waveform of variation represented by a curve shown in Figure 23.

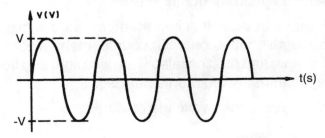

Figure 23

Electronics for the Electrician

This is a the curve of a trigonometric function called sine and it shows that the voltage across of the poles of an AC generator changes slowly from the positive to the negative passing across the zero point, pumping the electrons forward and backward through a circuit.

When working with electronics, the electrician can find both DC circuits and AC circuits. The circuits powered from batteries or cells are DC circuits. The stages of these circuits are powered from DC sources but they can have some parts where AC currents are found. In general, AC currents ranging from low to high frequencies found in these circuits are called signals. They are used to transport some kind of information like sounds or images.

For the circuits powered from AC power, there are three situations to take into consideration:

1. Some use only AC in their circuits such as the ones used to control AC appliances (motors, lamps, etc).
2. Some use AC in the input but convert it to DC and then DC is used in almost all the internal circuits.
3. Some can use both AC and DC depending on the stage. And, in many stages we can find 60 Hz AC currents and signals ranging from a few hertz to hundreds or thousands of megahertz.

This means that the electrician working with electronic equipment must be prepared to find both AC and DC voltages independent of the used power supply.

h) Sound and Electromagnetic Waves

Many electronic circuits are designed to operate with sounds or electromagnetic waves (radio waves). Telephones, wireless alarms, electronic doorbells, and remote control systems are examples of electronic devices that work with sounds and/or electromagnetic waves.

When working with these devices it is important for the electrician to know something more about the "vibrations." Of course, many electric devices such as doorbells, lamps, and buzzers also work with light and sound, but in a more simple way than many electronic devices.

First of all, we must separate the two kinds of vibrations:

Section I—The Differences Between Electricity and Electronics

1. Sounds

When a metal bar is struck the vibrations produce disturbances in the surrounding air. These disturbances are caused by compression and decompression waves that propagate through the air in all directions, as shown in Figure 24.

When reaching our ears these waves give us the sensation of sound. But, to be heard as sounds the disturbances or waves must have some special characteristics. First, the human ear can only perceive vibrations inside a limited range of frequencies or "speed" of vibrations. We can hear only the vibrations in the range of about 16 to 18,000 Hertz (vibrations per second). For example, if the bar vibrates at 30,000 Hz it produces sounds that are beyond the audible limit—ultrasonic vibrations. Many animals (dogs, bats, and dolphins) can hear ultrasonic sounds.

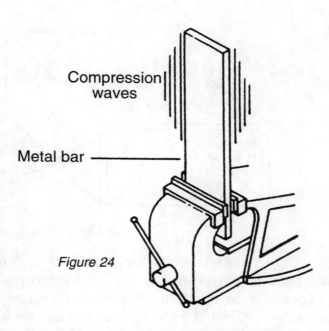

Figure 24

Another important characteristic of the sounds is that the human ear can differentiate them by their frequency. Low-frequency sounds, between 16 and 500 Hz, are perceived as bass. High-frequency sounds are perceived as treble. Our ears are sensitive enough to differentiate two sounds if their frequencies are separated of 1/16 of their value. It is because of this that we can differentiate two adjacent musical notes by the number of vibrations or frequency.

Also important is the way the metal bar is put into vibration. The natural way of vibration for a "perfect body" like a tuning fork is the one that produces a sine sound wave as shown in Figure 25.

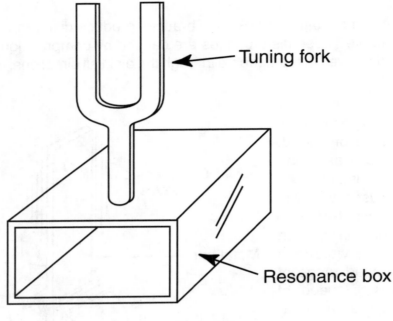

Figure 25

But, if the vibrations are produced by "imperfect bodies" or bodies with special formats such as musical instruments that have resonance boxes or special sizes and formats, the waveform can be different. The different waveforms give the sound a characteristic referred to as timbre or color. Timbre allows us to differentiate between the same tone or musical note produced by a violin and a piano; they have the same frequency but different timbre.

Sounds need a material to support their propagation. What this means is that they can't propagate in a vacuum. In the air, the speed of a sound wave is about 340 m/s.

2. Radio or Electromagnetic Waves

It was in 1865 that the physicist and astronomer James Clerk Maxwell published a theory explaining the existence of electromagnetic fields or electromagnetic waves.

To explain this, begin with the idea that a standard electric charge is surrounded by an electric field and a moving electric charge is surrounded by a magnetic field as shown in Figure 26.

Section 1—The Differences Between Electricity and Electronics

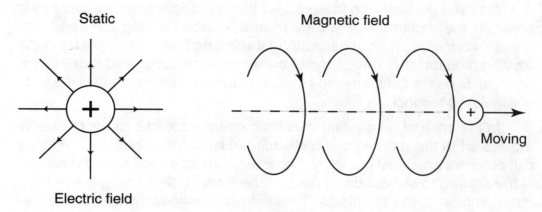

Figure 26

From this, a simple imaginary experiment can be done to show Maxwell's theory using a row of swinging metal balls (often sold as desktop toys) as shown in Figure 27. If metal balls are placed end to end in a row and then one end is struck by another moving ball, as shown in (a), the motion of the striking ball is transmitted through entire row. There is little loss as the motion is passed to the ball at the other end of the row as shown in (b).

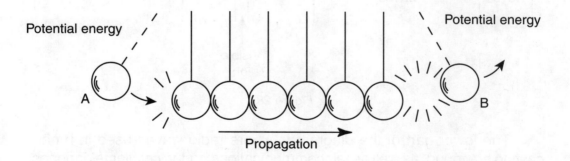

Kinetic energy in propagation

Figure 27

Maxwell showed with this process that a detectable time delay exists between the departure of energy from a source (striking ball) and that energy's arrival to the last ball, during the interval the energy exists in the moving medium (the row of metal balls). He also concluded that the energy exists half in the form of the motion of the medium and half in the form of an elastic reliance.

In the case of an oscillating electric charge, the two forms of energy are stored in the magnetic and electric fields, correspondent to what we call potential and kinetic energy. So, when a charge oscillates, the energy is transferred to surrounding space in the form of electromagnetic waves that combine in the two fields. These electromagnetic waves can travel through space with the speed of light or 300,000 km/s (as we will see, light is an electromagnetic wave).

Electromagnetic waves are present in our world in many forms. Because electromagnetic waves can be produced in any frequency from zero to infinity, they are represented by a continuum or spectrum as shown in Figure 28.

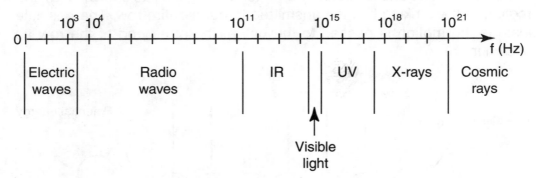

Figure 28

The lower part of the diagram is where radio waves used in broadcasting are found, as well as telecommunications, radar, cellular telephones, TV, remote controls, etc. A little further up is the IR or Infrared part of the spectrum. This corresponds to heat waves. Radiation in this part of the spectrum is due to thermal agitation of an object's atoms. The next segment consists of visible light with the colors, followed by UV waves (ultraviolet radiation). At the upper end of the spectrum are X-rays, gamma rays,

Section I—The Differences Between Electricity and Electronics

and cosmic rays with very high frequency and energy. For the electrician who wants to work with electronics, the most important parts are those corresponding to radio waves, infrared, and visible light.

Two important quantities are associated with electromagnetic waves—wavelength and frequency. Any electromagnetic wave can be represented by sine curves corresponding to the electric and the magnetic fields as shown in Figure 29.

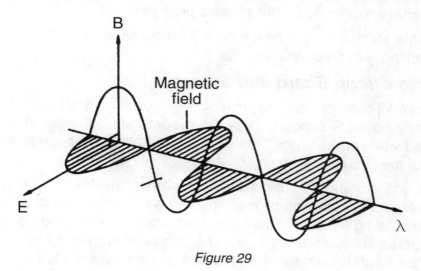

Figure 29

The frequency is the number of waves produced in each second and is measured in hertz (Hz). As high frequencies are common in electronic circuits, the use of the prefixes kilo (k), mega (M), giga (G), and tera (T) are common. When starting from the source, the distance traveled during one cycle can be defined as the wavelength (λ). This distance depends on the speed of the electromagnetic wave that, in a vacuum, is 300,000,000 m/s (or 300,000 km/s). So, to calculate the wavelength of an electromagnetic wave, divide 300,000,000 by the frequency in hertz.

If, for example, the wavelength of a 100 MHz (100,000,000 Hz) signal of an FM radio broadcast station is:

λ = 300,000,000/100,000,000 = 3 meters

Knowing the wavelength of a radio wave is very important since it determines the characteristic of the antenna used to transmit or receive it.

Electronics for the Electrician

The sizes of the elements of any radio antenna depend on the wavelength of the signal it is designed to use.

The fundamentals of electricity are the same that command electronic circuit operations. In electronic circuits, AC and DC currents pass through components and use all the common effects as you most likely studied before. Light, heat, magnetic fields, motion, images, and sounds are generated by devices that operate based on principles found in any electric devices or are explained by the physics principles.

For the electrician, the next step is to know what tools must be used when working with electronic circuits.

i) Printed Circuit Board and Solder

The small components used in electronic equipment can't stand alone without any physical support. They need some kind of support to keep them fixed while at the same time providing an electrical connection with the rest of the circuit.

Opening any piece of electronic equipment, the electrician will find that the electronic components are mounted on a special board of fiber or another insulating material. This support or chassis for the components is called a printed circuit board (PCB). The board, as shown in Figure 30, is made of an insulating material where copper strips are printed on one or both sides.

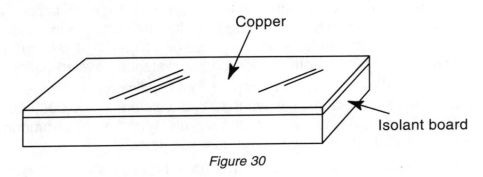

Figure 30

The strips act as wires, conducting the currents from one component to another. The pattern of the strips is determined by the function of the circuit. All the strips are planned before the manufacturing process to provide the necessary connections between components in the desired func-

Section I—The Differences Between Electricity and Electronics

tion. This means that a printed circuit board produced to receive components that form a radio can't be used to mount TV circuits or any other equipment.

As shown in Figure 31, the small components are soldered onto the board so their terminals make contact with the copper strips.

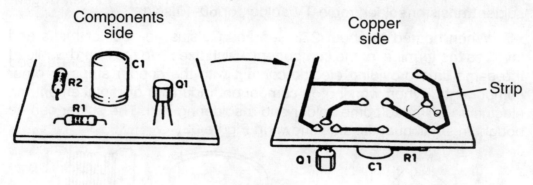

Figure 31

In some cases, when the components are very small as in surface-mounted devices (SMD), they can be placed on the board at the same side of the strips as shown in Figure 32.

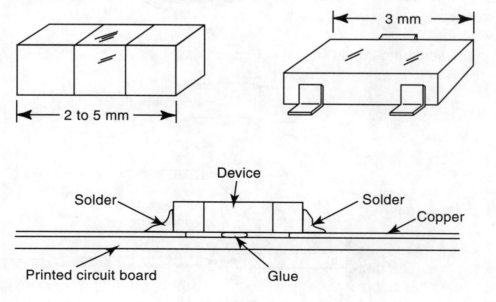

Figure 32

Removing and reinstalling components on a printed circuit board is a delicate operation that an electrician who wants to work with electronics, to repair and mount circuits, must know. Special techniques are required for this work because the components are so small and extremely sensitive to heat. The solder used in electronics is an alloy formed of 60 percent tin and 40 percent lead with some rosin. It is common to call this kind of solder transistor solder, radio-TV solder, or 60-40 solder.

When heated to about 273 degrees Celsius, the solder melts and involves the terminal of the component which fixes it to the board, while at the same time providing electric contact with the copper strips or other components. When working on component mounting or replacement, an electrician will need some solder and a soldering iron. The solder can be bought in small quantities as shown in Figure 33.

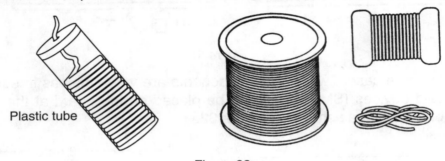

Figure 33

A soldering iron is shown in Figure 34.

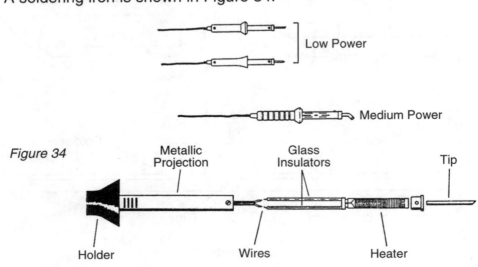

Section I—The Differences Between Electricity and Electronics

A 25- to 40-watt soldering iron with a shiny tip is recommended when working with small electronic components found in circuits. Of course, the electrician can also have a heavy-duty soldering iron to remove or place larger components like those found in some electric and electronic applications. Radio Shack has soldering irons suitable for this task as well as the next task:

- Dual Wattage Pencil Iron, 15 to 30 W (Radio Shack 64-2060)
- 15 W Pencil (Radio Shack 64-2052)
- 25 W Pencil (Radio Shack 64-2072)

Soldering is a simple operation and many electricians are familiar with this kind of work. But, electronic devices are very special and care should be taken when soldering to avoid damaging them. Many electronic devices are easily damaged by excess heat or the incorrect soldering procedure. The basic procedures to solder electronic components (remove or install on a PCB) are as follows:

- Plug in the soldering iron and heat for at least five minutes. This is long enough to bring the tip to the correct temperature for good soldering operation.
- Touch the soldering iron to the work area, allow a very short time for the connection to heat up, and then touch the solder to the connection, not the iron, as shown in Figure 35.

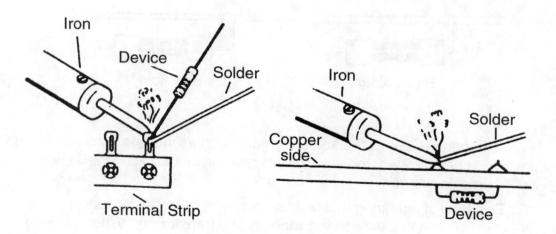

Figure 35

- You will notice that when the solder melts it penetrates every part of the solder joint.
- Remove the iron and do not move the joint until it has had time to cool. It is easy to see when the joint is cool. A peculiar haze will pass over the hot metal after which, the joint is cool enough and strong enough to withstand movement.

Figure 36 shows a perfect solder joint and some solder joints with problems. One of the principal causes of problems in electronic equipment is "cool solder." The solder seems to involve the component, but no electric contact is established because the joint was not heated enough to penetrate the metal creating an isolated layer of moisture or oxide to form between them.

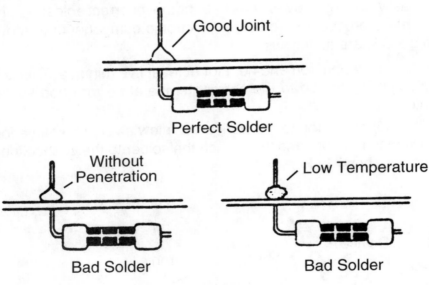

Figure 36

Although the use of the soldering gun, such as the one shown in Figure 37, is not forbidden, there are many cases in electronic works in which it is not recommended.

This type of soldering tool has its resistive tip heated by a strong current coming from the low-voltage winding of a transformer. When working with sensitive components such as integrated circuits, transistors, and others during the soldering operation, this current can be enough to burn them.

Section I—The Differences Between Electricity and Electronics

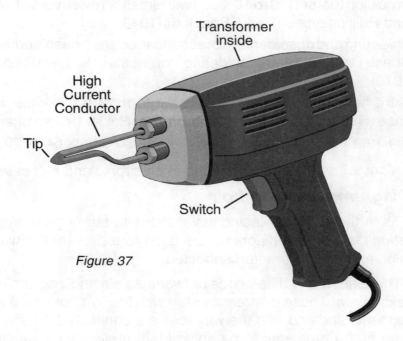

Figure 37

j) Other Tools

The soldering iron isn't the only special tool needed by electricians who intend to work with electronic circuits.

When working with electronic parts, many of the tools used in electric installations or automotive electricity are suitable. On the other hand, many electronic components are very small and delicate and need special tools and care. The use of improper tools when working with these components is often the cause of damage. If an electrician intends to work with electronic circuits, some special tools are needed. I suggest having at least some of these tools:

- Cutting pliers or diagonals (often called dykes) in sizes from 4 to 6 inches.
- Chain-nose or needle-nosed pliers with very narrow tips in sizes from 4 to 5 inches.
- Two or more screwdrivers sized between 2 and 8 inches. (Seven-piece set: Radio Shack 64-1823).
- Crimping tools—stripper and cutter for 10 to 22 wire gauges (Radio Shack 64-2129).

Electronics for the Electrician

- Precision tool set (10 to 16 tools) with small screwdrivers of hex, common, and Phillips types (Radio Shack 64-1948).
- Soldering and desoldering accessories such as a desoldering bulb (Radio Shack 64-2086) and a soldering iron holder/cleaner (Radio Shack 64-2078).
- Extra hands to hold the work in the form of a mini-vise with vacuum base (Radio Shack 64-2094) or a Project Holder (Radio Shack 64-2093).
- Mini hand drill (five-piece hand drill, Radio Shack 64-1779).

Many other tools can be found in electronic and tool catalogs.

k) Diagrams and Symbols

As in the case of electric installations in buildings or automotive installations, schematic diagrams are used to represent the way the many parts of equipment are interconnected.

There are many differences between an electric diagram and an electronic diagram. These differences start with the symbols used to represent components and end with the way they are connected. These differences are one of the problems to be solved by an electrician wanting to know something about electronics. He needs to be acquainted with new symbols since there are many electronics components represented by symbols that the electrician doesn't know. Learning the symbols used to represent each component and what they do in a circuit are the next steps in this book.

Let's begin with the schematic diagrams.

Figure 38 is a diagram of a simple piece of electronic equipment: a DC power supply similar to those used to power small calculators, CD players, and other appliances from the AC power line.

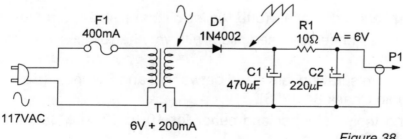

Figure 38

Section I—The Differences Between Electricity and Electronics

This schematic diagram represents all components by their symbols, and in many cases, the identification, values, and other important information are given. At the side of the symbol of each component is the identification number. This is important since it can help the technician find the component on the PCB or inside the equipment.

For example, all the resistors are identified by the letter R followed by the number of the device in the equipment. This means that many resistors can be found in a device identified by R1, R2, R3, etc. Capacitors are usually noted by the use of the letter C. The capacitors of a circuit are numbered starting from C1, C2, C3, etc. Transistors can be identified by the letters Q, T, or TR. They can be represented as Q1, T1, or TR1.

In many cases, a second number can identify the "block" or a "stage" in which the component is placed. So, the resistors of the first stage can begin at R101 and the resistors in the second stage from R201. Near the identification and the symbol we can also find the value or type of the component.

Resistors have the value of resistance to the side, such as R1, 1000 ohms or 1k. If it is a transistor, you might find 2N3906, meaning that when used, the transistor must be replaced with a 2N3906. The usual identification number for transistors begins with a 2N, but many manufacturers have started using letters signifying their names such as TIP (Texas Instrument) and MPS or MM (Motorola). There is also a European code that uses the letters BC or BD in the identification of devices, and a Japanese configuration that uses the letters 2SB, 2SC, or 2SD to indicate transistors.

Depending on the circuit, other important information can be found in the schematic diagram. For example, the voltage at different points in the circuit can be found.

In the example presented, it is indicated that, between A and the ground (normally taken by reference or 0 V), the measurement is 6 V when a multimeter is used to take the voltage (Figure 39).

Another important piece of information found in a diagram is the waveform of the signal present in a point.

As we said, electronic circuits don't operate only with DC voltages but also with AC voltages in a large range of frequencies and waveforms. These AC voltages or signals are not only sine signals, but can have several

Electronics for the Electrician

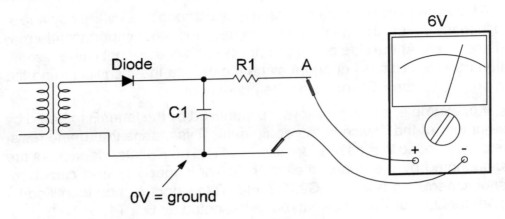

Figure 39

waveforms and amplitudes. Oscilloscopes are used to observe waveforms of a signal in a circuit. Figure 40 shows a common oscilloscope used on electronic workbenches for troubleshooting and diagnostics.

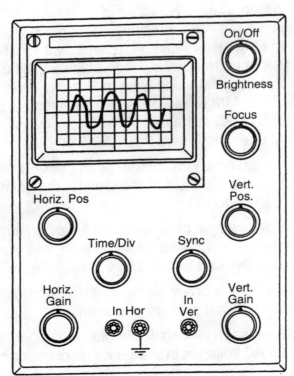

Figure 40

Section I—The Differences Between Electricity and Electronics

By observing the waveform of a signal at a point in a circuit a technician can determine what is wrong. The oscilloscope is one of the most powerful tools in the electronic lab. In the schematic diagram example there is the indication of a waveform at point B.

Also found on a schematic are procedures for installation, diagnostics, equivalence, etc. Diagrams for many types of equipment can be found with their manuals, but this is not common. A solution for the electrician who has problems with a piece of equipment and needs the schematic diagram is to purchase one. Sams Photofact is the most important supplier of schematic diagrams for commercial electronic equipment. You can access the company's Web site at www.samswebsite.com.

Section 2

Electronic Components

The parts found in electronic equipment are very different from the ones the electrician finds in electric installations of buildings and cars. Knowing how these electronic components work, how they are used, and how to read and interpret their specifications is an important item to everyone who intends to be an electronic technician. It is also very important to know how these parts can be tested.

So, as a brief introduction of electronics to the electrician, the principal electronic components are described. These are found in some equipment for use in domestic installations and also the automobile. From this description, the electrician will be able to recognize these components when working with electronic equipment. This knowledge is also very important as a starting point for a basic course in electronics.

Wires

Wiring material is used to connect the circuit parts, devices, equipment, etc., electrically. There are various types of wiring material for use in

Electronics for the Electrician

electronic applications. The wires used in electronic equipment have some differences from the ones found in electric installations. The current flow, in general, is lower in electronic appliances, which need thinner wires, but some devices use special wires. Basically the electrician will find three types of wires in electronic work:

A. Cable—formed by a core of several copper conductors and covered by plastic insulation.
B. Rigid—formed by a core with only one conductor covered by plastic insulation.
C. Bare—formed by a copper conductor without cover (insulation).
D. Screened and Coaxial—formed by a core with one or more plastic covered conductors. These conductors are enveloped by a copper screen. Over the entire cable there is a plastic cover.

The plastic-covered wires can be found in several forms including single wires, twisted wires, flat cable, etc. Figure 41 shows some types of wire used in common wiring of electronic circuits and parts.

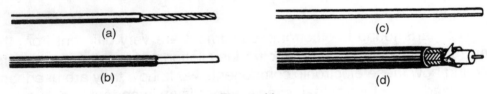

Figure 41

AWG (American Wire Gauge) is a special type of enameled wire that is used in some electronic (and also electric) components such as coils, relays, solenoids, motors, magnetic sensors, loudspeakers, earphones, and many others. AWG is a standard U.S. set of nonferrous wire conductor sizes. The gauge is referring to the diameter. This standard is mostly applied to copper household wiring and telephone wiring. AWG is sometimes known as Brown and Sharpe (B&S) Wire Gauge, also.

AWG enameled wire is formed by a rigid copper wire insulated by a cover of enamel. The diameter of the wire determines the amount of current it can conduct. The diameter of a wire can be expressed in millimeters, mils, or by an AWG number. The higher the AWG number of a wire, the smaller the diameter. Table 1-2 shows the characteristics of the AWG wires used in electronics.

Section II—Electronic Components

AWG No.	Dia.	Cross Sec. (mm²)	Resistance (ohms/km)	AWG No.	Dia.	Cross Sec. (mm²)	Resistance (ohms/km)
0000	11.86	107.2	0.158	21	0.7230	0.41	41.46
000	10.40	85.3	0.197	22	0.6438	0.33	51.5
00	9.226	67.43	0.252	23	0.5733	0.26	56.4
0	8.252	53.48	0.317	24	0.5106	0.20	85.0
1	7.348	42.41	0.40	25	0.4547	0.16	106.2
2	6.544	33.63	0.50	26	0.4049	0.13	130.7
3	5.827	26.67	0.63	27	0.3606	0.10	170.0
4	5.189	21.15	0.80	28	0.3211	0.08	212.5
5	4.621	16.77	1.01	29	0.2859	0.064	265.6
6	4.115	13.30	1.27	30	0.2546	0.051	333.3
7	3.665	10.55	1.70	31	0.2268	0.040	425.0
8	3.264	8.36	2.03	32	0.2019	0.032	531.2
9	2.906	6.63	2.56	33	0.1798	0.0254	669.3
10	2.588	5.26	3.23	34	0.1601	0.0201	845.8
11	2.305	4.17	4.07	35	0.1426	0.0159	1,069
12	2.053	3.31	5.13	36	0.1270	0.0127	1,339
13	1.828	2.63	6.49	37	0.1131	0.0100	1,700
14	1.628	2.08	8.17	38	0.1007	0.0079	2,152
15	1.450	1.65	10.3	39	0.0897	0.0063	2,669
16	1.291	1.31	12.9	40	0.0799	0.0050	3,400
17	1.150	1.04	16.34	41	0.0711	0.0040	4,250
18	1.024	0.82	20.73	42	0.0633	0.0032	5,312
19	0.9116	0.65	26.15	43	0.0564	0.0025	6,800
20	0.8118	0.52	32.69	44	0.0503	0.0020	8,500

Table 1-2
Standard annealed copper wire (AWG & B&S)

Fuses

Fuses are the protection elements of a circuit.

The whole purpose of a fuse is to switch off the current in the circuit before anything other than the fuse can be harmed when a problem occurs. They operate based on the old principle of a chain being only as strong as the weakest link. It is easy to see that the best way to protect everything else was to make one link deliberately weakest so it would fail first.

The name fuse comes from fusible link, as suggested by the previous analogy. A fuse is designed to conduct its rated current and act as a conductor. When the current increases beyond the rated value the fuse melts, acting as an SPST switch and opening the circuit.

Symbols and types

Figure 42 shows the symbols used in electronics diagrams to represent a fuse and also the types of common fuses found in electronic equipment. These fuses are basically formed by a piece of wire in which thickness is the determining factor of the melting current.

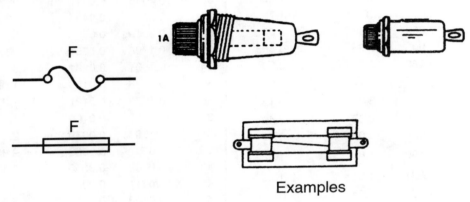

Examples

Figure 42

Where they are found and how they are used

The fuse is wired in series with the device or circuit to be protected as shown in Figure 43. It is usual in some electronic equipment to have several fuses, each one protecting one stage or part in such a manner that if one stage fails and the corresponding fuse melts, the other stages can continue their normal operation unless they are interdependent.

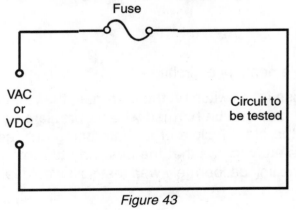

Figure 43

Specifications

The most important specification of a fuse is the melting

Section II—Electronic Components

current or the current that causes the fuse to burn. This nominal current can be specified in miliamperes or amperes according to the value and application.

When replacing a fuse, never use a unit with current specification larger than the original. The specification of current must be the same. If a larger fuse is used and a failure occurs in the circuit, increasing the current to a dangerous value, the fuse may not be the first component to burn out. It may be some component of the circuit or possibly many of them that will burn.

Other specifications of a fuse to consider are how fast it burns when problems occur and the highest operation voltage. In some electronic circuits the fuse is a small piece of bare wire placed between two terminals. The wire gauge is chosen to melt with the desired current.

How to Test a Fuse

A good fuse must present a low electric resistance. To know if a fuse is in good condition, test its continuity—whether it can conduct the electric current. A high resistance means a burned fuse (open circuit). A good fuse measures a resistance near zero when tested. A multimeter can be used for this task.

Switches

In electronic appliances the electrician will find many types of switches. As in any other electrical application, the switches are used to control the current flow throughout the circuits or parts of them.

The difference between the switches used in common electric installations and cars is not important. In electronic equipment the switches, in many cases, are used to control many currents at the same time. The design also depends the applications.

Symbols and Types

Figure 44 shows the symbols and types of the main types of switches found in electronic circuits. The symbol indicates what the switch makes. For instance, (a) shows an SPST (Single-Pole Single-Throw), which is a

Electronics for the Electrician

switch that controls the current flow in only one circuit (single-pole). As the name suggests, this switch controls one circuit turning the current on and off.

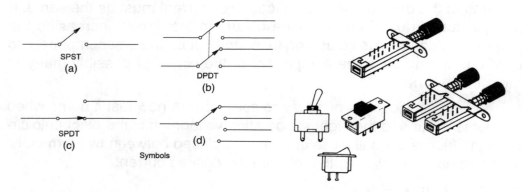

Figure 44

A DPDT (Double-Pole Double-Throw) is shown in (b). This switch can control the current in two independent circuits (double-pole) transferring it to one or another circuit (double throw). The switches also are different in the way they are operated. A circuit can have slide switches (b), toggle switches (c), and rotary switches (d).

Some special switches are also shown in Figure 44, such as thumbwheel types, pushbuttons, and subminiatures that are suitable for installation in PCBs (Printed Circuit Boards), etc.

Where they are found

Switches are found in any application where current must be controlled. Every piece of electronic equipment needs at least one switch for its operation—the on/off switch.

Specifications

The switches are identified by the number of poles, the number of positions, the maximum operating voltage, and current.

The poles and positions are indicated by symbols or letters. These are the main types:

SPST—Single-Pole Single-Throw
SPDT—Single-Pole Double-Throw
DPDT—Double-Pole Double-Throw

Maximum Operating Current and Voltage

The maximum operating current is indicated in amperes (A). Never use a switch in a circuit with a current larger than the maximum recommended. The contacts can heat causing problems. The maximum voltage is indicated in volts (V).

How to Test

An open switch must present a very high resistance (near infinite) and when closed a very low resistance (near zero). To test a switch, take a simple continuity measurement using a multimeter in the lowest resistance scale. It is important, in some cases, to make a visual inspection: a switch with contact problems due to excess current can present some visible signs such as deformations or dark areas in the contact.

Cells and Batteries

Cells are primary chemical energy sources for much electronic equipment. The cell is basically formed by two different metals immersed in some kind of chemical substance. The cells used to power electronic equipment can be found in a large collection of types, shapes, and sizes. The type, size, and shape are determined by the energy requirements of the application, the length of time the battery must source energy, and the overall size of the equipment to be powered. Many types of technology are used by the manufacturers to provide all the batteries found in electronic equipment. The technology involved is with the materials used inside the cell and its form as they are installed.

The most common cell is the carbon-zinc dry cell or Leclanché cell that is found in many battery sizes (button, AA, AAA, C, and D). This cell consists of one electrode of carbon and one of zinc with an electrolyte of ammonium chloride. Actually, the carbon electrode is composed of a mixture of manganese dioxide, carbon, and solid ammonium chloride. Most of the energy produced by this cell comes from decomposition of the manganese dioxide.

A variation of this type of cell is the alkaline cell. In the alkaline-manganese cell the chemical reaction is the same as in the carbon-zinc cell, but it takes place in an alkaline medium rather than the acid mixture used in the carbon-zinc cell. This kind of cell can provide up to twice the current output with a much longer shelf life.

These common cells are not rechargeable as they operate based in a nonreversible chemical reaction, thus the name primary cells. On the other hand, rechargeable or storage cells are called secondary cells—such as the ni-cad (nickel-cadmium) types.

These cells can be recharged by a current flowing in the reverse direction than the current source flows when they are in operation. Special devices, such as the one shown in Figure 45, are used to source this current.

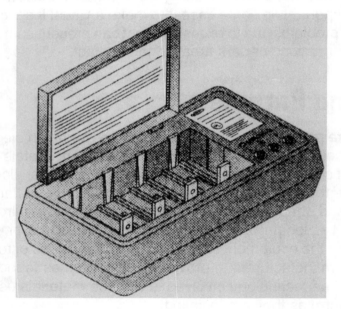

Figure 45

Common cells source voltages between 1.2 V and 1.5 V.

When many cells are associated in series we call the association a battery. The most common type of battery found in electronic applications is the 9 V battery. It is formed by the association of 6 single 1.5 V flat cells.

Symbol and Types

Figure 46 shows the most common types of cells and batteries found in electronic applications. Each type can be found in dry, alkaline, ni-cad, and some other versions based on the application.

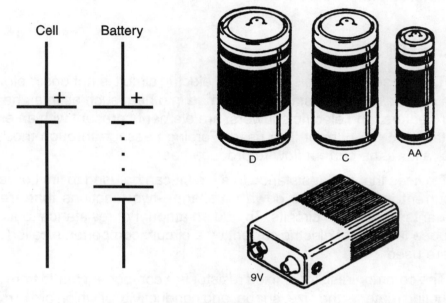

Figure 46

Specifications

A. The first and principal specification of a cell or battery is the type and size. Standard designations are groups of letters to indicate the format and size (AA, A, C, D, etc.).

B. The second specification, voltage, can be omitted because it is included in the type.

C. Depending on the application it is important to know how many hours the battery can source a determined amount of energy to a device. This is indicated by the battery or cell Ah or mAh (amperes x hour or milliamperes x hour). For example, a battery with a 100 mAh specification can source 10 mA of current to a circuit during 10 hours or 1 mA during 100 hours.

Testing

Many battery testers are available in specialized stores. But, a simple multimeter can also be used to make this test. Many of them have a battery test function. The measure of the voltage of a cell is not a conclusive way to test a battery or cell.

Resistors

The presence of resistance in an electric circuit is not desirable. Only when electric power must be transformed into heat such as in air heaters, water heaters, and electric showers, are elements present with an electric resistance to the current flow used. Forcing passage through those elements allows the current flow to produce heat.

Devices that add resistance to a circuit can be used to limit or reduce the current flow or to step down a voltage—two functions that are very important in electronic circuits. To add an amount of resistance to a circuit or reduce the flow of electric current by a circuit, components called resistors are used.

The common resistor or fixed resistor is a component made from some conducting material the size, shape, and conductivity of which are arranged to determine the amount of electrical resistance needed. The amount of resistance is measured in ohms (Ω).

Inside electronic equipment are resistors of many types and sizes with resistance values ranging from less than 1 ohm to many million ohms. The resistors are classified into fixed or variable resistors. Let's look at fixed resistors, also called resistors.

Resistors are found in many sizes and shapes according to the amount of resistance they present, some special performance feature they must have, and the amount of heat generated when working.

Symbols and Types

Figure 47 shows the symbols used to represent a resistor and the types of resistors commonly found in electronic circuits. The symbol shown in (a) was adopted by the American system of symbols. In European equipment the schematics or circuit drawings represent the resistor as shown in (b).

Section II—Electronic Components

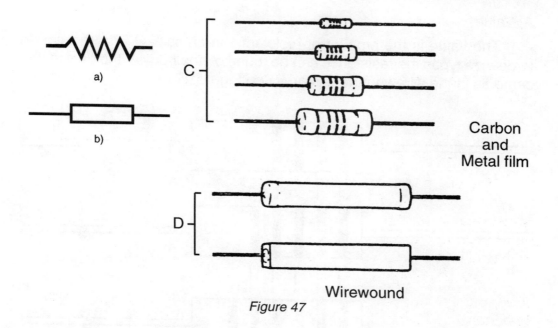

Figure 47

The most commonly used resistors are carbon composition and carbon film types, as shown in Figure 47 (c) where the resistive element is a thin layer of carbon strips deposited onto a ceramic rod. Another type is the metal film resistor or film resistor, with the same external aspect as the carbon type. In this type, metal is evaporated onto ceramic rods forming a thin film. These resistors are cheap and can be easily fabricated in a large range of values. They are intended for applications in low-power or small-current circuits. The disadvantage of carbon resistors is that they are noisy.

When desiring resistors that can operate with large currents, the wirewound type shown in Figure 47 (d) is preferred. Those resistors are formed by turns of metal wires (alloys such as nickel, chromium, iron, silver, etc.) onto a ceramic rod. The material, gauge, and number of turns determine the amount of resistance, the size, and how much heat they can handle.

Specifications

When replacing or testing any electronic device it is necessary to know its specifications. The main specifications for resistors are:

Electronics for the Electrician

A. Value

The value is the amount of resistance in ohms (Ω) of a resistor. In large-size types the resistance can be found on the body of the device, the same as in the wirewound resistors (See Figure 47).

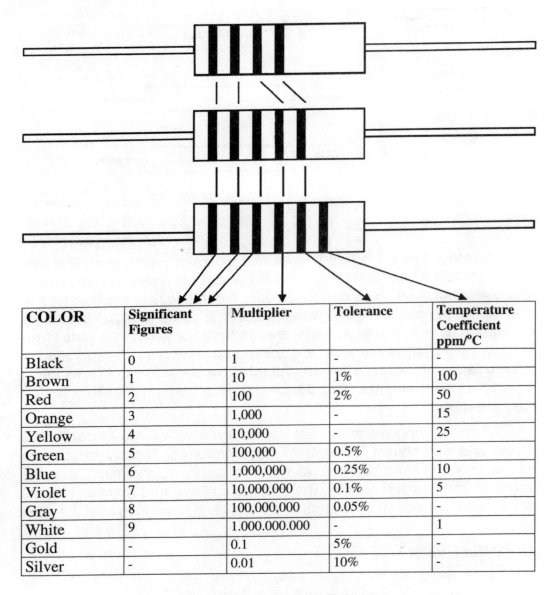

COLOR	Significant Figures	Multiplier	Tolerance	Temperature Coefficient ppm/°C
Black	0	1	-	-
Brown	1	10	1%	100
Red	2	100	2%	50
Orange	3	1,000	-	15
Yellow	4	10,000	-	25
Green	5	100,000	0.5%	-
Blue	6	1,000,000	0.25%	10
Violet	7	10,000,000	0.1%	5
Gray	8	100,000,000	0.05%	-
White	9	1.000.000.000	-	1
Gold	-	0.1	5%	-
Silver	-	0.01	10%	-

Table 1-3 Color Code for Resistors

Section II—Electronic Components

But, in smaller types such as carbon and metal film types as well as SMDs, there isn't enough space to display the resistance value. In this case, and also to make the manufacturing process a bit easier, a color code is used. Color bands or stripes are printed on the resistor body. Accordingly, the position in the equipment is part of the meaning of a band. Combining the two, the resistance value of any resistor can be found. In addition to the value in ohms, the color bands also indicate the tolerance of the component.

Table 1-3 indicates how the value and tolerance of a resistor is read using the color code.

Notes on Table 1-3
1. If the resistor is a three-band type, the tolerance can be assumed as 20%
2. If the resistor is a three- or four-band type the temperature coefficient is not indicated.

Example—A resistor has colored bands in this sequence: red, violet, orange, and silver. (Read the colored bands from the end to the center.) Bands 1 and 2 form the number 27. Band 3 indicates that we must multiply this value by 1,000. The resistance is 27,000 ohms. Band 4 indicates that the tolerance is 10%.

B. Dissipation

The size and the material the resistor is made of determines how much heat it can produce without burning. Based on the application the resistors are to be used for, resistors must be chosen in different sizes or dissipation capacities.

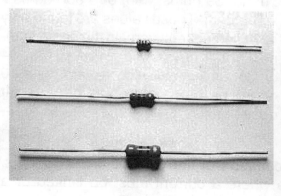

Figure 48

The dissipation or power rate of a resistor is specified in watts (W). Small resistors (metal film or carbon) have dissipation in the range between 1/8 and 2 W. Wirewound types can be found in dissipation starting from 1 or 2 W to more than 200 W. Figure 48 shows the differences in size between carbon resistors according to their dissipation.

(Figure 48) Carbon resistors with different dissipations are shown. From the smaller to the greater the dissipations are 1/8, ¼, and 1/2 W.

C. Tolerance

It is impossible to manufacture a resistor with an exact value of resistance. This is not necessary as the electronic circuits are designed to operate within a certain band of value of components and voltages. The difference between the value specified for a component and the real value or measured value is called tolerance.

Resistors can be found in tolerance ranges from 1 percent to 20 percent. This means that a resistor indicated as 1000 ohms/10 percent unit can really present resistances between 900 and 1100 ohms. As a consequence, it is not necessary make resistors with all values between 900 and 1100 ohms, as the 1000 ohms resistor covers that range. So, the electrician will be surprised when observing that the resistors are found in few common values, according to their tolerances. The basic values are determined by a commercial series of values.

Table 1-4 shows that resistors of 47 ohms, 470 ohms, 4700 ohms, 47,000, 470,000 x 20 percent and many other values in multiples of 10 can be found, but not 39 ohms, 390, or 3,900 ohms x 20 percent. These values are found only in 10 percent resistors. In the same manner, you'll never find a 3.4 ohms, 34 ohms, 340 ohms, etc., resistor with 10 or 20 percent tolerance because it doesn't exist.

20% SERIES OR E-6	10% SERIES OR E-12
10	10
	12
15	15
	18
22	22
	27
33	33
	39
47	47
	56
68	68
	82
100	100

Table 1-4

Where Resistors are Used

The resistors are the most common of the electronic components. They are used in all applications where a current must be limited or a voltage steps down.

Section II—Electronic Components

In electronic circuits the main application for resistors is in bias or stepping down a voltage. Many elements of a circuit must operate with voltages lower than supplied by the power supply. This means that resistors must be used to step down the voltage to the desired value. These resistors are called bias resistors.

Testing

When heated by excess current, the resistors can open or burn. Although a dark area in the body of the component is an evident sign that it is damaged, in some cases the component doesn't present any other visible signs of problems.

Resistors are tested using the multimeter (VOM). Desolder one terminal of the component and take it off of the printed circuit board or out of the circuit and measure the resistance. Don't measure it while still in the circuit because the multimeter can indicate the resistance of the circuit, not that of the component.

Figure 49 shows how to do this test.

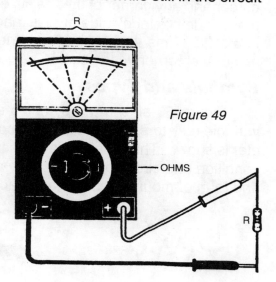

Figure 49

Remember to consider the tolerance when reading the value displayed by the multimeter and comparing it with the value displayed by the component. When replacing a resistor use a unit of the same type and same value. The dissipation, in some cases, can be higher if enough space is available for its placement.

Variable Resistors

In many applications something in a circuit can be controlled or adjusted by changing a resistance. You can increase the volume of an amplifier or the brightness of a lamp by altering the current in a circuit. The common way to alter currents and voltages in circuit is altering a resis-

Electronics for the Electrician

tance placed in the circuit for this task. The principal component that can be used to make this happen is the variable resistor. The electrician will find variable resistors in a large number of functions in electric and electronic appliances.

As the name indicates, a variable resistor is a device presenting a resistance that can be changed. Variable resistors are found in different sizes and aspects. There are two types. One allows the resistance value to be changed easily and often, like the volume or tone adjustment of an audio amplifier. The other is the semifixed resistor that ordinarily doesn't change the resistance value. It is used to adjust the operational condition of the electronic circuit.

Basically, variable resistors are formed by a resistive material (metal or carbon) with a cursor sliding onto it. When the resistance between the cursor and any corner of the resistive material changes normally, between 0 and a determined value of resistance, it is called nominal resistance. So in a 1000 ohms variable resistor, the resistance between the cursor and any of the corners can assume values between 0 and 1000 ohms.

Symbols and Types

Figure 50 shows the symbols and types of the two basic types of variable resistors. In (a), a trimmer potentiometer or adjustable potentiometer is shown. This small variable resistor is used to adjust the operational condition of a circuit one time. They are found inside the equipment. Size and form of mounting can change according to the application.

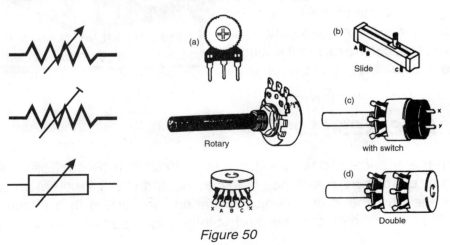

Figure 50

Section II—Electronic Components

In (b) are potentiometers (slide and rotary) that are used to control many functions of equipment. The slide potentiometers are common in audio equipment and the rotary types are found in TV, radio, dimmers, etc. Many potentiometers are double (d) as those used to control the volume in both channels of a stereo amplifier. Others (c) can include switches to turn on and off an appliance.

Specifications

A. Value or nominal resistance

The most important specification in a variable resistor (potentiometer or trimmer potentiometer) is the nominal resistance. This resistance is indicated in ohms (Ω) and represents the maximum value of resistance assumed by the component. For example, a trimmer potentiometer preset with 10,000 ohms or 10kΩ can be adjusted to any resistance between 0 and 10,000 ohms.

B. Variation of resistance

In some applications the way the resistance changes when the potentiometer is adjusted is important. There are basically two types of variation of resistance curves: linear and logarithmic.

Where they are found

The electrician can find trimmer potentiometers and potentiometers in many types of electronic equipment. In every application where a function of a circuit must be adjusted or controlled by changing current in a circuit, a variable resistor can be used.

An example of a common use is in a dimmer where the amount of current in a circuit determining the brightness of a lamp is controlled by a potentiometer. Timers also use a potentiometer to adjust the time.

Testing and replacing

Trimmer potentiometers and potentiometers can burn or present many functional problems. One of the problems is the cursor contact can fail. Another common problem is caused by noise generated when the cursor slides on the resistive element. This occurs mainly in audio equipment. This problem is the cause of the scratching noise produced when the volume control of a radio or amplifier is adjusted.

A basic test can be done with a multimeter to measure the resistance of the resistive element (between the extreme terminals in potentiometers). Another test is done by sliding the cursor and listening to hear if noise is produced (if a multimeter doesn't indicate abrupt jumps in the resistance change).

Incandescent Lamps

Small incandescent lamps or bulbs are used in electronic equipment as panel indicators, flashlights, or even as active elements of a circuit. The nonlinear characteristic of the filament (voltage vs. current) makes incandescent lamps ideal for use as a current regulator in some applications.

Incandescent lamps are formed by a tungsten filament placed inside a glass enclosure where the oxygen is removed. When the electric current heats the filament to a very high temperature, it produces light. The metal of the filament doesn't burn because there is no oxygen to be a source of a combustion chemical reaction.

Electronic equipment usually uses small incandescent lamps rated for low voltage in the range between 1.5 and 12 V. This is an important difference from the lamps found in electric applications that are primarily used to illumination purposes and are powered directly from the AC power-line voltage.

Symbols and Types

Figure 51a shows the symbols used to represent small incandescent lamps. In (a), the American symbol is shown and in (b), the symbol found in some diagrams from European manufacturers is shown. Figure 51b illustrates the common types used as panel or instrument indicators and for other functions.

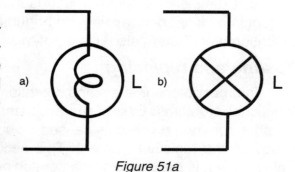

Figure 51a

Section II—Electronic Components

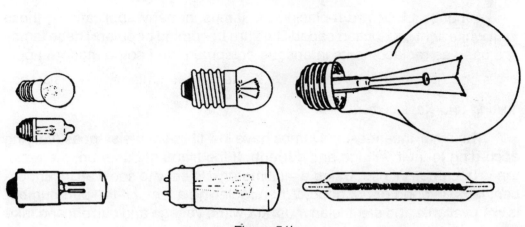

Figure 51b

Observe that the base of the lamp changes in format or size according to the application. The main types are the bayonet base, telephone slide, flange, and threaded.

Specifications

A. Nominal voltage

This is the operational voltage of the lamp. Common types are indicated to operate in the range between 1.5 and 12 V.

B. Nominal current

This is the current flowing through the filament when the nominal voltage is applied. Common types can be specified to operate with currents in the range between 0.01 A (10 mA) and 1 A.

C. Type number

Many catalogs indicate the type of a lamp by a part number giving tables where the electric and mechanical specifications are placed. Types such as "47" are popular. This is a lamp rated to 6.3 V x 150 mA using a bayonet base.

Where they are found

The incandescent lamps can be found basically in all applications where illumination of a panel, instrument, or control is needed. In many pieces of equipment, small incandescent lamps are placed inside the panel instruments.

Although LEDs are replacing the lamps in many applications, those with white light production capabilities are becoming popular. These lamps are used as replacements in antique equipment and some modern applications.

Testing and Replacement

The small incandescent lamps have low filament resistances (varying according to their voltage and current). If the filament burns up, the resistance becomes infinite. Using a multimeter, it is easy to see if an incandescent lamp is good or not good. When replacing a lamp, if the part number is not available, the electrician must know the voltage and current and take care to choose one with the same base format.

Neon Lamps

Neon lamps are used in many electronic circuits as indicators or important active elements because their electric characteristics (negative resistance) make them ideal as trigger devices in timing circuits or waveform generators. The common neon lamp is formed by a small glass bulb filled with neon gas. Two electrodes are placed inside without one touching the other. When the voltage across the gas increases reaching the triggering point (between 60 and 80 V), the gas ionizes and becomes a conductor. The electric resistance falls allowing the circulation of a current. With the ionization, the gas emits an orange light.

Symbols and Types

Figure 52 shows the symbol used to represent the neon lamp and types of this device.

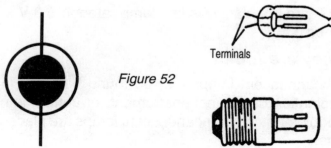

Figure 52

Section II—Electronic Components

Specifications

The main specification of a neon lamp is the trigger voltage. Normally it is in the range between 60 and 80 volts. The manufacturers indicate neon lamps by a part number such as NE-2H or HE-51. Sometimes it is important to know the type of lamp and if it does or does not have a current-limiting resistor inside.

Where they are found and how they are used

Neon lamps can be used both as indicators and elements of an active circuit. As indicators, they can be found in the input of equipment monitoring the AC power line voltage as shown in Figure 53.

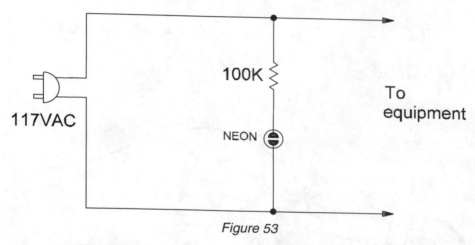

Figure 53

The electrician will find that in circuits like this, the lamp has a series current-limiting resistor (typically between 47k and 1M). Some types of lamps such as the NE-2 have this resistor inside for 120 V use. Others need an external resistor.

Another application is an oscillator for generating signals as in the configuration shown in Figure 54. Depending on C and R values, the circuit can generate sawtooth audio signals or make the lamp flash.

An interesting application of the neon lamp is as a phase indicator as shown in Figure 55. If placed in the phase terminal of an AC power line, the lamp will glow. The electrician will not feel any shock because the series resistor limits the current to a very low value, not enough to excite the nervous cells, giving a shock.

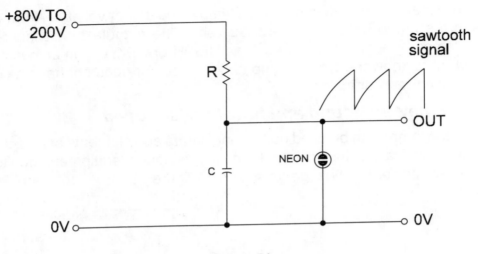

Figure 54

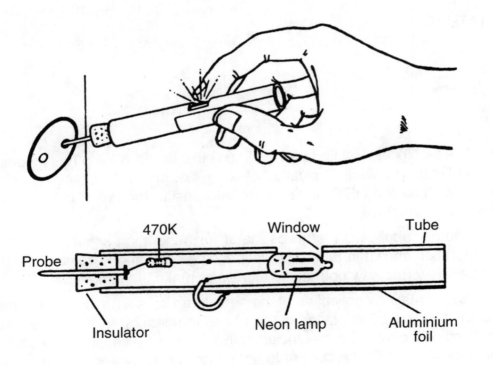

Figure 55

Section II—Electronic Components

Testing

The simplest way to test a neon lamp is to plug it into the AC power line using a 220k ohms series resistor.

WARNING: *Do NOT plug the lamp directly into the AC power line. Without the current limiting resistor it can explode!*

Light-Dependent Resistor or CdS Cells

Light-dependent resistors (LDR) or CdS cells or CdS elements are light-sensitive resistors presenting a resistance value that changes with the light. The resistance between the terminals of these components falls when the amount of light falling onto a sensitive surface increases. In the dark, the resistance is high and falls under the sunlight. The LDRs are used as light sensors in many electronic applications.

They are formed by a cadmium sulfide surface—the sensitive element of the device. This surface and its base are placed inside a transparent plastic enclosure. The size and the format of the device depend on the application.

Symbols and Type

Figure 56 shows the common symbols used to represent this component and Figure 57 shows what they look like.

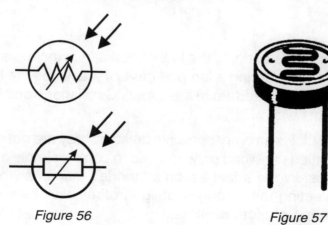

Figure 56   Figure 57

Specifications

The LDRs can by specified by the diameter (size) or by a manufacturer number. When replacing LDRs, types of the same size are normally interchangeable.

Where they are found

The electrician will find LDRs in many electric devices related to domestic applications. They are used as sensors in alarms, automatic lights, emergency lights, and many other circuits. In a typical application the LDR is placed in a way that it receives only the ambient light as shown in Figure 58.

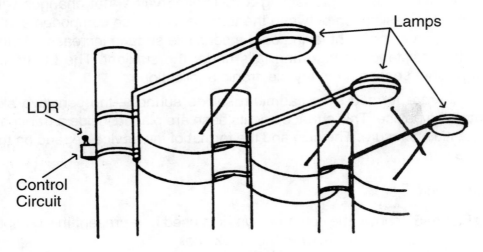

Figure 58

When the light falls at dusk, the LDR's resistance increases acting on the electronic circuit triggering a lamp. If the LDR receives the lamp's light a feedback process can interfere in the circuit. Oscillations and instabilities can be induced this way.

Although the LDRs are very sensitive devices, they are not fast enough for certain applications and can only be used to control simple circuits from slow light changes. When a fast action is needed, such as when reading bar codes or detecting fast changes of light, other devices are preferred like photodiodes and phototransistors.

Section II—Electronic Components

In many applications the amount of light on the LDR or CdS cell can be increased with the use of a convergent lens. The use of a convergent lens, as shown in Figure 59, is important to help the device to detect light from smaller angles.

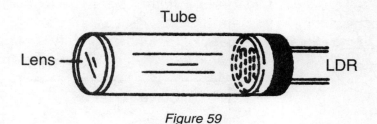

Figure 59

Testing

A simple multimeter (digital or analog) can be used to test any LDR. Put the multimeter on a resistance scale and plug the probes into the LDR. Passing your hand in front of it (to cut the ambient light falling in the sensitive surface) you'll observe the resistance changes in the multimeter's scale.

NTC/PTC

Negative Temperature Coefficient resistors (NTC) or Positive Temperature Coefficient (PTC) resistors (also called thermistors or thermally sensitive resistor) are components having a resistance that changes with their temperature. In the NTC the resistance falls when the temperature increases and in the PTC the resistance increases with the temperature.

They are formed by materials (alloys and mixtures) that have special thermal properties suitable to the task.

Symbols, Types, and Characteristic Curve

Figure 60 shows the symbols used to represent these components, types, and the characteristic curve.

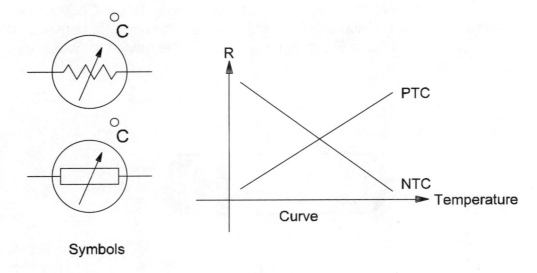

Figure 60

The size of the component determines how fast it is when altering its resistance during a temperature change. This is an important factor in some applications and is referred to as readiness. The format and size depend on the application.

Specifications

The NTCs and PTCs can both be specified by a manufacturer or part number and a nominal resistance, or a resistance in the ambient temperature (normally 20 degrees Celsius). Also refer to Figure 60 for the characteristic curve, a graphic representation of the changes of resistance with the temperature.

Where they are found

The electrician will find NTCs and PTCs in many electronic circuits used in electric installations or even in electronic appliances. Some fire alarms and ambient temperature controllers used with air conditioners, as well as many other devices, use one of these components as a sensor. They are installed in a place where the temperature can be detected. The signal is then sent to a remote device that acts on the element to be controlled (air conditioner, alarm, fan, relay, etc).

Section II—Electronic Components

Testing

The NTCs and PTCs can burn or break. In this case, the resistance will increase and it can open the circuit. The simplest test consists of taking the resistance measurement using a common multimeter.

Moisture can alter the nominal resistance of high-resistance NTC or PTC inducing a circuit into a wrong operation. This is important to be considered if the NTC or PTC is used as a sensor and not protected from moisture or ambient action.

VDRs

Voltage Dependent Resistors (VDR) or Zinc-Oxide Varistors and Metal Oxide Varistors or Transient Surge Absorbers are devices having a resistance that changes with the applied voltage. They are nonlinear resistors and can be used in circuit protection and many other applications such as power supplies, microprocessor protection, surge protection for consumer and industrial electronics, and many other applications.

The VDR is a device that becomes a conductor when the voltage applied to it crosses a determined value.

Symbols and types

The symbols and examples of common VDRs are shown in Figure 61. The size of the VDR determines the amount of energy it can absorb when becoming conductor.

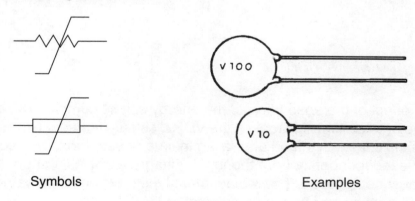

Symbols   Examples

Figure 61

Electronics for the Electrician

Specifications

The main specification of a VDR is the breakdown voltage. It is the voltage at which the device becomes conductive, passing from a high-resistance state to a very low-resistance state. This component can be found with voltage specifications between 18 and 1800 volts.

Where they are found and how they are used

The electrician will find VDRs at the input of many electronic devices operating as a protective element. In fact, when voltage spikes appear in the AC power line, if their value is high enough to make the VDR conductive, the snap action of this component puts a short in the spike. So, the high voltage that could damage the equipment is shunted to the ground, or the energy of the spike absorbed by the VDR.

Computer outlet sockets with surge protection use this device as a main element in the protection circuit. Outlets for computers with surge protection such as the one shown in Figure 62 use VDRs as protection elements.

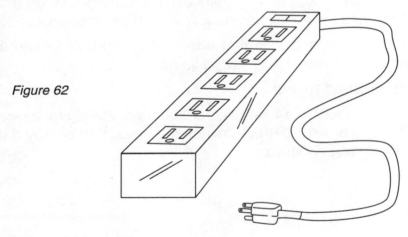

Figure 62

Testing

The effect of the absorption of the energy spikes coming from an AC power line or telephone line makes the VDR dissipate their energy as heat. This effect is cumulative, and after a lag in time of use, the VDR becomes conductive independently from the input voltage acting as a short. In this case, it must be replaced. The replacement type must have the same voltage as the original and be the same size.

The multimeter can be used for testing. A good VDR presents a high resistance above a predetermined hundred thousand ohms.

Capacitors

Capacitors have the function of storing up electrical energy. The basic capacitors are devices formed by a piece of insulating material between two pieces of conductive material. When DC voltage is applied to the capacitor, the electricity carried by the electric charge is stored up to each electrode. The capacitor energy lies in the electrical field formed between the pieces of metal or electrodes called plates. The insulator material is also referred to as dielectric. In electronic equipment, capacitors are used in many important functions. They can be found in many formats, sizes, and with different dielectric materials. The name of a capacitor is given by the material used as the dielectric. For example, a capacitor that uses ceramic as its dielectric is a ceramic capacitor, and a capacitor using mica as its dielectric is a mica capacitor.

Symbols and Types

The types of commonly used capacitors and the symbols used to represent them are shown in Figure 63.

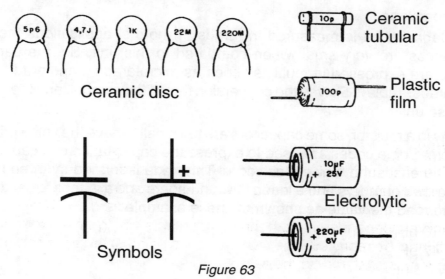

Figure 63

Some differences in representations are also found between diagrams as the European symbol used to represent capacitors is different than the American.

About the types, notice the differences in the manufacturing processes that result in tubular types (where the metal plates and dielectric form a coil) and the plate (where the dielectric is a planar piece).

The differences in types are very important as each capacitor has specific applications according to some electric properties of the material used and the format. For example, ceramic capacitors are suitable for high-frequency applications although polyester film is not. The plastic film capacitors such as polyester and polypropylene are used in low and medium frequency applications.

It is also important to know that electrolytic capacitors are polarized components. This means that they have definite positive and negative electrodes or plates. They can't be inverted when used in a circuit because this can cause the breakdown of the material used as dielectric conductor. In some cases, if a large current flows through the component during a fail, an explosion can occur.

Specifications

The electrician, when replacing a capacitor, must be aware of these specifications:

A. Capacitance

Capacitance is measured in farads. In normal applications the capacitances are very small when compared to this unit. So, the capacitances are expressed in multiples such as microfarads, nanofarads, and picofarads. Table 1-5 gives the conversion factor, the symbol, and the value of these units.

As in a resistor, some capacitors are too small to have the value printed on them. Some codes are used to express the capacitance of a capacitor. One of these is the "three-digit code." This code is formed by three numbers, or two numbers and a letter. If two numbers and a letter are used, it is easy to read the value as shown by these examples:

10n = 10 nF
47p = 47 pF
3.3n = 3.3 nF

Section II—Electronic Components

Unit	Symbol	Value in Farads (F)
Microfarad	μF	$0.000\ 001 = 10^{-6}$ F
Nanofarad	nF	$0.000\ 000\ 001 = 10^{-9}$ F
Picofarad	pF	$0.000\ 000\ 000\ 001 = 10^{-12}$ F

To convert	In	Multiply by:
Microfarads	Nanofarads	1,000
Microfarads	Picofarads	1,000,000
Nanofarads	Picofarads	1,000
Nanofarads	Microfarads	0.001
Picofarads	Microfarads	0.000 001
Picofarads	Nanofarads	0.001

Table 1-5

If three numbers are used, the first two digits show the first and second figures of the capacitance and the third becomes a multiplier which determines how many zeros are to be added to the capacity as in these examples.

102 = 10 00 = 1000 pF

473 = 47 000 = 47 kpF or 47 nF (k = 1000)

Under 100 pF only two digits are displayed: 27 is 27 pF.

Figure 63, where many types of capacitors are shown, gives some samples with this kind of value indication.

Working Voltage

The capacitance of a capacitor depends on the distance between plates and the nature of the material used as a dielectric. The thicker the insulator in a capacitor, the greater its volume, but the smaller its capacitance. Otherwise, if the insulator is very thin, it can't isolate high voltages. Also, the physical construction of a capacitor is important to indicate the highest voltage that can be applied between plates. This is the maximum operating voltage, sometimes expressed as a WVDC (Working Voltage DC).

Types with working voltages in the range from very low voltages to more than 1kV are found in electronic equipment.

Type

The principal types of capacitors found in electronic circuits are:

- Electrolytic capacitors (Electrochemical type)

This type of capacitor uses aluminum as the electrode and a thin oxidization membrane as the dielectric. This layer is so thin that high capacitances can be achieved. But, since the layer is formed in an electrolytic process (thus the name), they are polarized. The pole is identified by a mark in the component (see Figure 64). Electrolytic capacitors have a pole indication that must be observed when installing them. Electrolytic capacitors are found in a capacitance range from 1 uF to more than 220,000 uF and are used in DC and low-frequency circuits.

- Tantalum capacitors

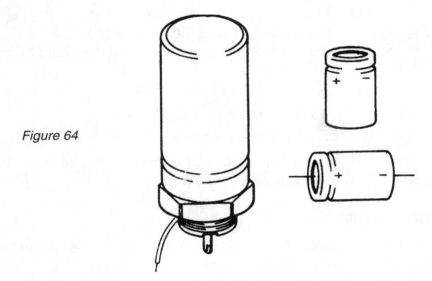

Figure 64

This is a special type of electrolytic capacitor using the material called tantalum for the electrode. Since tantalum oxide has greater dielectric constant when compared with aluminum, tantalum capacitors can be made with very large capacitances and small sizes. Even this capacitor there is polarity. It usually uses the symbol (+) to show the positive lead. Tantalum capacitors are found in a range of values starting from 0.1 uF.

- Ceramic capacitors

Section II—Electronic Components

Special ceramics, such as titanium acid barium, are used as a dielectric in this kind of capacitor. The electric properties of the dielectric make this capacitor useful in high-frequency applications. The most common type is the disk style. The capacitance of ceramic capacitors is comparatively small. They are typically found in values between 1 pF and 470 nF.

- Polystyrene film capacitor

This capacitor is included in the family of the plastic film capacitors. A thin film of polystyrene is used as a dielectric in this capacitor. Since the plates and the dielectric form thin foils that are wound in a coil, the component presents some inductance. Because of this factor, this capacitor is not recommended for use in high-frequency circuits.

- Polyester film capacitor

Another type of plastic film capacitor is the polyester film capacitor. This kind of capacitor is recommended for high-frequency applications and can be found in a range from very low picofarads to 10 uF or more.

- Other capacitors

Other capacitors in this component family are polypropylene capacitors used in applications where low tolerances are needed, mica capacitors used in applications where high stability is needed, and metalized polyester film capacitors used in miniaturization applications.

In antique equipment other types such as oil/paper capacitors can be found. These capacitors use a paper infused in oil or dry paper as a dielectric. When restoring a piece of antique equipment where a damaged oil or paper capacitor is found, it can be replaced by an equivalent plastic film capacitor.

Where they are found

The electrician is familiar with capacitors since they are used with motors and in other electrical applications. However, the capacitors found in electronic equipment are found in a much larger assortment of types, values, and sizes as well as in every part.

Capacitors have special electric properties that make them very important to the performance of all electronic circuits. They can be used to measure time intervals, to perform calculus in computers, and to separate

AC signals from DC signals in audio and high-frequency circuits. They are also used as filters and for energy storage in many applications.

Testing

When connecting the probes of an analog meter to the terminal of an electrolytic capacitor (10 uF or greater) a current flow is registered by the needle's movement. This current flows while the electric charge is storing up. The current stops flowing when the charge finishes storing. The oscillating movement of a needle in a multimeter can be used to test an electrolytic capacitor. Use an intermediate resistance scale for this task.

Once charged, capacitors are like open circuits. When perfect, they can't let the current pass through them. This means that good capacitors have a very high resistance that can be measured with the aid of any common analog or digital multimeter. If a capacitor presents low resistances (under a hundred thousand ohms) we say that it presents losses or is shorted and must be replaced.

Variable Capacitors

In some applications a capacitance presented by a capacitor must be changed. This is the case in a radio receiver where the station tuned is determined by a capacitance of a resonant circuit. Every time we need to change the station the capacitance in that circuit must be altered. This task needs the aid of a special capacitor. The capacitor that can change the capacity is called a variable capacitor.

Like variable resistors, variable capacitors are found in several types. All of them are manufactured starting from the same principles: a moving plate can slide in front of a fixed plate with an insulator between them, or a moving plate can change the distance from a fixed plate also with a dielectric between them.

Two basic variable capacitor types are found in electronic applications: the trimmer, one that can be adjusted by a screwdriver, and the variable capacitor, or varicon, that is adjusted by a button placed in its axis.

Section II—Electronic Components

Symbols and Views

Figure 65 gives the symbols used to represent variable capacitors and views of the most common types.

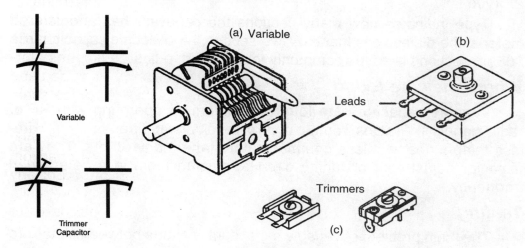

Figure 65

The variable capacitor shown in (a) is found in antique devices such as radios and tuners. The dielectric is the air gap between the moving plates and the fixed plates. In (b) are variable capacitors like those found in modern small transistor radios. In this type the dielectric is a polyester foil between the moving plates and the fixed plates.

The trimmer capacitors can be found with plastic, ceramic, and mica as the dielectric. Adjustment is made by a screw that moves one of the plates as shown in Figure 65 (c).

Specifications

Two main specifications are important when choosing variable or trimmer capacitors.

1. Capacitance

For variable capacitors, normally maximum capacitance is shown as nominal capacitance. Thus, a 120 pF capacitor is a capacitor that can change capacitance between a smaller value and 120 pF. However, even when totally open the capacitance isn't zero, which is normal, to indicate the minimum capacitance. For trimmer capacitors the capacitance range is given. For example, a 2-20 pF trimmer capacitor can be adjusted to present any value between 2 and 20 pF.

2. Type

Type indicates how many sections the capacitor has and of what material the dielectric is made. A two-section, air-dielectric capacitor with 365 pF sections is a type commonly found in old radios.

Where they are found

Variable capacitors are found in equipment operating with RF or radio signals like radios, remote control receivers and transmitters, wireless telephones, wireless communicators, and many others. They are used for adjustment or tuning circuits, placing them in a determined frequency.

Testing

The main problem of variable capacitors is shorts between plates. To test this, measure the resistance between the fixed and moving plates (the component must be disconnected from the circuit for this test). The resistance must be infinite in any position of the adjustment to indicate a good device.

Coils or Inductors

Coils, chokes, or inductors are components made of a certain number of turns of wire on a form. They are used to provide an inductance wherever it may be required. Inside the form is air (no core or air core) or some magnetic material such as ferrite (iron dust), iron, etc. The coils present opposition to current changes characterized by an electric quantity called inductance. The inductance is measured in henry (H) but, the fractions—millihenry (mH) and microhenry (µH)—are widely used in practical components. The number of turns and the gauge of wire determine the inductance of an inductor or coil.

Three major classes of coils or fixed inductors are available. They are the filter choke, audio frequency inductor, and the RF inductor or choke. (RF means radio frequency).

Section II—Electronic Components

Symbols and types

Figure 66 shows some types of coils or inductors the electrician can find in electronic equipment.

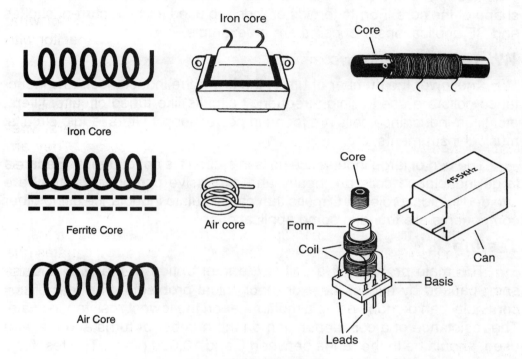

Figure 66

The presence of a continuous line in the symbol indicates an iron core. The presence of an intermittent line in the symbol indicates a ferrite core. If the line is absent the coil is an air core type.

Adjustable coils can also be found in some applications. These coils have an adjustable core that can be slid or screwed in their form.

Specifications

A. Inductance

The inductance of a coil is expressed in henry (H) or its fractions. Low values in the range of microhenry are found in high-frequency applications. Values in the range of millihenry and henry are found in low-frequency circuits and as filter chokes.

B. Core

The nature of the core is an important specification for a coil. Low-frequency and DC coils (filters and audio) use iron cores. High-frequency coils use ferrite and air cores. It is also important to indicate the size and shape of the core. Ferrite toroids or rods are used for high-current chokes and RF applications. Pot cores are an example.

Where they are found

The lower the number of turns, the lower the inductance. Low-inductance coils are used in high-frequency circuits like tuned circuits, filters, etc. High-inductance coils are found in power supply filters, audio circuits, musical instruments, etc.

The use of large inductance coils in a circuit is not common. Because large inductance coils are heavy and expensive components, they are avoided in most projects. Circuits that can simulate inductance using other components are found in some applications.

Testing

The main problem of a coil is the interruption of the wire and the short caused by excess heat or mechanical problems. The test of coil continuity can be done using a multimeter on the lowest resistance scale. The resistance of a coil, depending on the number of turns and the wire used, should be in the range between 0 and 10,000 ohms. The test for a short is more difficult. Generally, a coil in short, when caused by excess current, presents burning signals such as dark spots and the smell of smoke.

Transformers

Transformers are components formed by two or more coils or inductors coupled so closely together that the magnetic field of one causes induction of current in the others. For the total coupling the coils are wound in a common form or core. When an alternating voltage is applied to one coil of a transformer (called the primary winding), this voltage induces alternating voltages in the other coils (secondary windings) depending on the ratio of turns of the two windings. For example, if a transformer has a

Section II—Electronic Components

primary winding with 200 turns and a secondary with 400 turns, and 110 VAC is applied to the primary, 220 VAC can be found in the secondary, as shown in Figure 67.

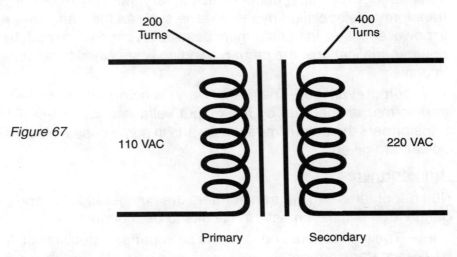

Figure 67

Transformers can be used to change voltages, currents, and impedances in an alternating-current circuit or high-frequency circuit. The transformers can be wound in forms with air cores, ferrite cores, or iron cores according to the application. Air and ferrite cores are used in high- and medium-frequency circuits and iron cores are used in medium- and low-frequency circuits.

Symbols and Types

Figure 68 shows the different symbols used to represent transformers and their types.

The presence of a material core (ferrite or iron) is represented by a continuous or intermittent line. If the line is absent it is an air core transformer.

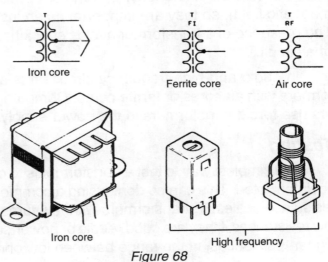

Figure 68

Specifications:

Audio and power supply transformers

- Input voltage or impedance—The primary winding or the input of a transformer is specified by the voltage such as the transformers found in power supplies. In transformers used as an impedance match in audio or other applications, the primary winding is specified by an impedance in ohms.
- The output voltage—The secondary winding of a power supply transformer is specified by the output voltage and current. For audio transformers the output impedance, and in some cases, also the output power, are given.

RF transformers

- Number of turns—In RF transformers the number of turns and the wire gauge are specified, mostly in circuits to be mounted.
- Core—The type of core and format is an important specification for high-frequency transformers.

Where they are found

The electrician will find transformers of many sizes and types in all electronic equipment. The most common types are the transformers used in the power supply of any equipment. The windings are isolated one from the other (the electric energy is transferred from one to the other by the magnetic field), so they are important as an insulation element in circuits. The presence of a transformer avoids shock hazards when touching parts of a circuit.

In audio and high-frequency circuits it is also possible to find transformers with air cores or ferrite cores. Computers and some audio amplifiers use toroid transformers in the power supply.

Testing

The simplest way to test a transformer is to check that the windings are not interrupted. This can be done using a common multimeter on the lowest resistance scale. RF transformers with few turns of wire must present very low resistances if they are good. Audio or power-supply transformers present higher resistances, in the range between low ohms and 10,000 ohms.

Section II—Electronic Components

Important: the measured resistance of the winding of any transformer isn't its impedance but the ohmic resistance of the wire used to wind it.

Relays

Relays are electromagnetic switches. They are formed by an electromagnet coupled to switch contacts so that the switch is operated automatically when the electromagnet is energized. Applying a voltage to the electromagnet allows the devices wired to the switch to be controlled. Relays are important as they can be used to control high power loads from weak signals or currents. Another advantage in the use of relays is that the contacts are isolated from the control circuit, meaning a relay can be used to control a high-voltage circuit from a low-voltage circuit. The disadvantage is that the coupling between the coil and the contacts is mechanical, limiting the speed of response.

Figure 69 shows the basic construction of a relay as found in many electronic applications.

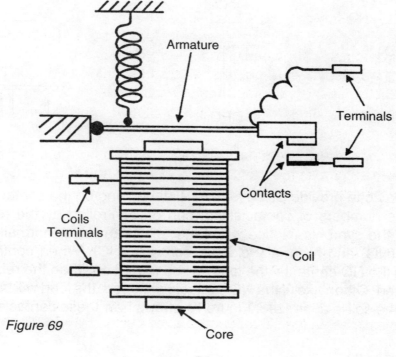

Figure 69

Electronics for the Electrician

The coil is formed by hundreds or thousands of turns of thin, enameled wire. The contacts are designed to control high currents based on the final use of the device.

Symbols and Views

Figure 70 shows the symbols and views of the relays used in common electronic applications.

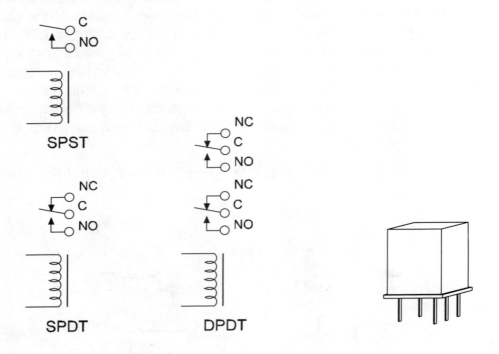

Figure 70

A relay can provide SPDT or DPDT depending on the contact placement. The numbers of contacts that can be controlled by the relay are shown in the symbols. Notice when the relay can have "normally open" contacts (NO) and "normally closed" contacts (NC). When controlling a load from the NO contacts, the load will be turned on when the relay's coil is energized. Otherwise, if the NC contacts are used, the load will be turned off when the coil is energized. Figure 71 shows how these contacts should be used.

Section II—Electronic Components

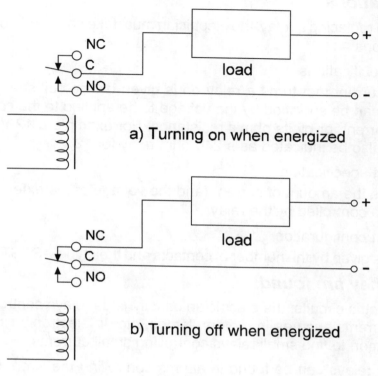

a) Turning on when energized

b) Turning off when energized

Figure 71

Figure 72 shows another important type of relay—the reed relay. This relay is formed by a coil involving a reed switch. The reed switch is formed by metal plates inside a glass enclosure that is filled with an inert gas. When a magnetic field acts on these plates, they touch one another, closing the circuit.

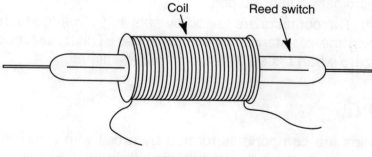

Figure 72

Specifications

When replacing a relay, the electrician must take care to observe these specifications:

A. Coil specifications

The current that turns a relay on is given by the coil specifications. The relay can be specified by the voltage to be applied to the coil in volts and the current, or by its ohmic resistance. For example a 12 V x 50 mA relay can also be indicated as a 240-ohm relay for 12 V.

B. Contact specification

This is the amount of current (and the voltage of the external circuit) that can be controlled by the relay.

C. Contact configuration

This is given by the number of contacts and their type—SPST or DPDT.

Where they are found

In electric circuits, the electrician usually finds relays controlling high electric currents such as in motors, lamps, etc. In electronic circuits it is most common to find small relays controlling small currents.

Small relays can be found in alarms controlling the siren, lamps, or other loads. In automatic doors, remote controls, and other applications, relays interface the control circuit with the controlled load.

Testing

Two kinds of tests can be done in a relay:

A. Coil: The state of the coil can be verified by a multimeter. A good coil presents resistances between some ohms and 10,000 ohms. This probe doesn't indicate a shorted coil.
B. Contacts: The contacts are tested by applying a voltage to the coil and measuring the resistance when they close. This resistance must be almost zero when tested with a common multimeter.

Solenoids

Solenoids are components formed by a coil with a moving core inside. When a current flows through the coil, the magnetic field acts on the core attracting it. Sliding inside the coil, the core can be used to move some mechanism.

Solenoids with different sizes and formats are found in electronic equipment. They are used to move mechanical parts of the appliances such as the tape-ejecting mechanisms in VCRs. Washing machines and dishwashers also use solenoids to control the flow of water.

Symbol and Views

The symbol of a solenoid and views of the common types found in electronic equipment are shown in Figure 73.

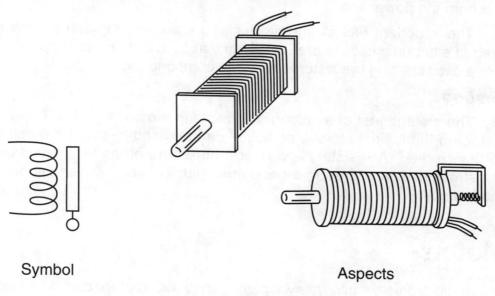

Symbol   Aspects

Figure 73

Small solenoids are formed by a coil with thousands of turns of a very thin, enameled wire. The core is a rod of iron that slides inside the coil. A spring and some mechanism can be found to return the core to the initial position when the current that causes the magnetic field disappears.

Specification

The main specification of a solenoid is the voltage that must be applied to energize it. As an important complement to this information, know the current drain when the nominal voltage is applied. In electronic circuits, solenoids with voltages ranging from some volts DC to AC power line voltage are found.

Where they are found

The electrician can find many types of solenoids in common electric appliances such as washers and electric doors. In electronic appliances, solenoids are found in items such as tape recorders, VCRs, CD players, and a large number of items with circuits where mechanisms must be controlled from electric currents. The solenoids found in these applications are generally very small and delicate. They are different from the ones found in electric appliances that are designed to operate with the AC voltage from the power line.

The important fact for the electrician to recognize is that solenoids used in electronic circuits are usually driven by DC currents supplied by active circuits using transistors and other components.

Testing

The electric test of a solenoid consists in the coil's probe. A multimeter, on the lowest resistance scale, can be used to see if the coil is not interrupted. A resistance varying from some ohms to a thousand ohms must be measured in a good solenoid. This test doesn't show a shorted coil.

Motors

In electronic circuits many types of small DC motors can be found. Basically a motor is a device that converts electric energy into motion. They can be used to move parts of a device from an in signal or control voltage. The electric motor is formed by one or more coils that produce a magnetic field when a current flows through them. The magnetic field acts on metal pieces or other coils causing a force that puts it in motion. In electronic applications, there are two main types of electric motors—DC motors and step motors.

These motors are formed by coils that, when energized, can place a shaft in a determined position. By energizing the coils in sequence, the shaft can run at a determined speed. DC motors produce continuous move-

Section II—Electronic Components

ment when a voltage is applied. Step motors can be used to place a piece in a predetermined position with great precision.

AC motors are more rare in electronic appliances. They are mainly high-power motors driving large mechanisms.

Symbol and Views

Figure 74 shows a parts view and symbols of some DC motors and step motors.

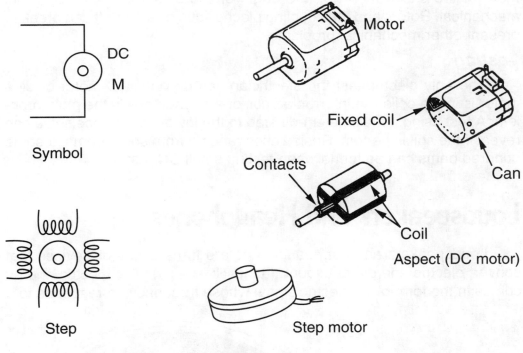

Figure 74

Specifications

DC motors are specified by the operational voltage and the nominal current. In these motors the current and the absorbed power are functions of the load. When running without a load the current is very low and the speed rises to the maximum value. When the load increases, the speed falls down and the drained current increases to the maximum. Another indication sometimes found is the speed given in rpm (rotations per minute) under nominal operation conditions.

The step motors are specified by the type (number of poles) and the voltage. The current drained is also an important specification for this kind of motor.

Where they are found

As stated, there are many electronic applications where motors can be found. Since they are delicate mechanisms, electricians will find that in more of the cases than not, the origin of a problem is not electric, but mechanical. Rotors and the coupling pieces such as the shaft can break or present other mechanical problems.

Testing

The only electric test the electrician can do on a DC motor or step motor is in the coils. An interrupted coil or a shorted coil is the main problem. A test using a multimeter adjusted to the lowest resistance scale can reveal if the coil is perfect. Resistances ranging from some ohms to some hundred ohms can be found when testing small DC motors.

Loudspeakers and Headphones

Loudspeakers and ear/headphones are transducers or devices that convert electric energy into sound (acoustic energy). The electric energy comes in the form of an electric signal whose frequency corresponds to a

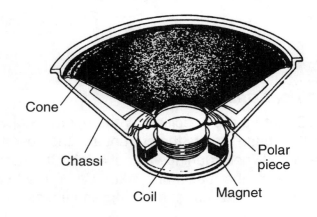

Figure 75

Section II—Electronic Components

sound. The moving coil loudspeaker (the most common in audio applications) is formed by a coil involving a polar metal piece where a strong magnetic field is produced by a magnet as shown in Figure 75.

The coil is fixed in a cardboard or plastic cone. When the signal is applied to the coil, its magnetic field interacts with the magnetic field produced in the polar piece, resulting in a force that moves the coil and the cone forward and backward. Compression and decompression waves are produced in the air surrounding the cone. These waves have a waveform corresponding to the original sound being reproduced.

In magnetic headphones small devices that operate according to the same principle can be found. Remember that both loudspeaker and headphones exist that operate from other principles such as the piezoelectric units. (See in the item on transducers for more information.)

Symbols and Types

Figure 76 shows the symbols and the types of some loudspeakers.

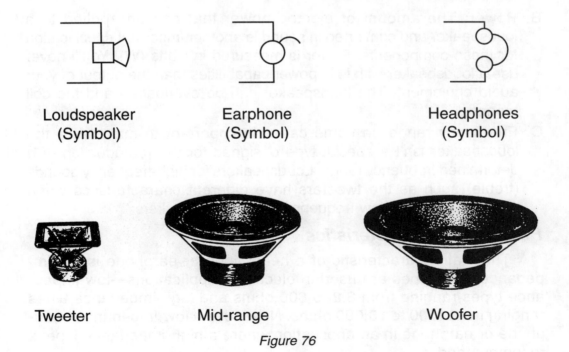

Figure 76

The larger the size of a loudspeaker, the easier the production of low-frequency sounds. This means that heavy magnets and large diameters are characteristics of low-frequency loudspeakers, also named bass speakers or woofers. Small loudspeakers are used for the reproduction of high-frequency sounds. These small speakers are called treble speakers or tweeters.

It is important to remember that many loudspeakers are designed to operate only inside enclosures.

Specifications

A. Impedance: When replacing a loudspeaker, next to the size, the impedance is the first electric specification to be observed. The impedance of a loudspeaker is measured in ohms and should not be confused with the ohmic resistance of the coil. Common types have impedances in the range between 3.2 and 16 ohms. Never use a loudspeaker with less impedance than the recommended; the output stage of the driver circuit can be forced and the components damaged.

B. Power: The amount of electric power that can be applied to a loudspeaker and converted in sound is another important specification for these components. Power is measured in watts (W). You'll never use a loudspeaker with less power capabilities than the output of your audio equipment. The loudspeaker can be overloaded and the coil burned.

C. Frequency range: In some cases is important to make sure the loudspeaker isn't a special type designed for the reproduction of a determined frequency range. Loudspeakers for high-frequency sounds (treble) such as the tweeters have different characteristics when compared with the low-frequency (bass) loudspeakers.

Headphone characteristics

The main characteristic of a headphone or earphone is the impedance. Two types are used in electronic applications—low-impedance types ranging from 3.2 to 600 ohms and high-impedance types ranging from 1000 to 10,000 ohms. Never use a low-impedance headphone or earphone in an application where a high-impedance type is recommended.

Testing

The electric test of a magnetic loudspeaker or headphone consists in the measure of the continuity of the coil. A common multimeter on the lowest resistance scale can be used for this task. The measured resistance of a good loudspeaker or headphone is very low, below 100 ohms for the low-impedance types. See that the measured resistance isn't the impedance but the ohmic resistance of the coil. Higher values, up to several thousand ohms, are common in high-impedance types.

Magnetic Transducers

Transducers are devices that convert one form of energy into another. A magnetic transducer transforms variable magnetic fields into electric currents. In the inverse mode, we can also consider a magnetic transducer to be any device that converts electric signals into magnetic fields. Many magnetic transducers can be found in electronic applications functioning as sensors. Basically they are formed by a coil that can be used to pick up any change in the magnetic field produced by a magnet under certain conditions, or a coil that produces a magnetic field from a current.

Loudspeakers and headphones are magnetic transducers, as explained previously, but are treated as a separate item because of their importance in electronics. Microphones and diskettes that read heads and position sensors are other examples of magnetic transducers used as sensors. The writing heads of a Winchester disk or disk drive are magnetic transducers operating in the inverse mode.

Symbols and Types

The symbols and types of some magnetic transducers are shown in Figure 77.

Specifications

Depending on the application, the transducer can show specifications as current, voltage, type of signal, or a part number. The most common is a part number specification.

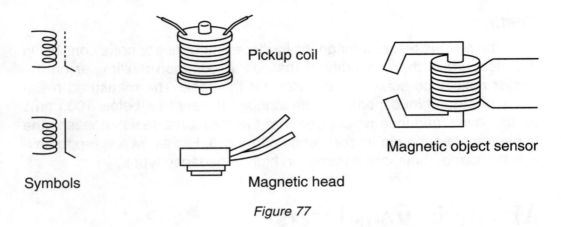

Figure 77

Where they are found

Pickup coils, sensors, microphones, Winchester disk, and disk drives are all examples of magnetic sensors.

Testing

A multimeter can be used to test the continuity of the coil. Magnetic transducers in general must present resistances ranging from less than 1 ohm to more than 1000 ohms based on the application. This test doesn't reveal other problems such as shorts or mechanical failures.

Piezoeletric Transducers

Some materials present piezoelectric properties due to their unusual crystal structure. Among these materials are quartz, Rochelle salt, and several kinds of barium and titanium ceramics. When a piezoelectric material is pressed, bent, or submitted to mechanical efforts or any change of form, electric charges appear in its surfaces. The same effect also works in the inverse mode; if we apply electric voltages to its surface, a mechanical deformation can occur. Some ceramic materials such as barium titanate and quartz crystals present piezoelectric properties.

These materials can be used to make piezoelectric transducers, piezoelectric sounders, piezoelectric loudspeakers and earphones, piezoelectric buzzers, and many other devices that convert electric energy into mechanical energy (sound or super sounds) or the opposite, such as microphones, vibration pickups, etc. Some components are very important in this family:

A. Quartz crystal: When excited by an electric signal quartz crystal tends to vibrate, producing signals of only one frequency. They can be used to determine the operational frequency of many circuits such as those in clocks, watches, computers, radio transmitters and receivers, and many others. The vibration frequency is established by the mechanical resonance of the crystalline plate.

B. Piezoelectric sound generators: A small piece of piezoelectric ceramic can be used to produce sounds in alarms, earphones, and high-frequency loudspeakers (tweeters). When an audio signal is applied, the ceramic vibrates producing sound waves. Ceramic transducers can also be used to produce supersonic sounds.

C. Piezoelectric high-voltage generators: Beating a piece of piezoelectric ceramic with a small, spring-triggered hammer, a ceramic piezoelectric transducer can generate very high-voltage pulses. The high voltage, up to 7000 volts, will produce a spark that is used in gas igniters and in other applications.

D. Piezoelectric microphones: Placing a diaphragm onto a piece of piezoelectric ceramic can be used to convert sounds or beats in electric signals. The sound waves beating the diaphragm make it press the ceramic piece producing the deformations needed to cause the electric signals to appear.

Symbol and Types

Figure 78 shows some piezoelectric transducers and their symbols.

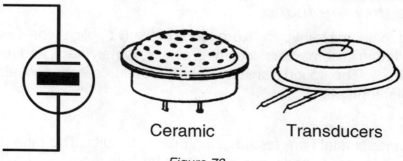

Ceramic   Transducers

Figure 78

Specifications

Normally the piezoelectric components are indicated by a part number determined by the manufacturer. The quartz crystals are indicated by the frequency. They are high-impedance devices because the ceramic and the quartz are good isolators. In some cases, a drive circuit can be installed inside the component, such as in piezoelectric buzzers that include an audio oscillator. Piezoelectric tweeters, for example, include a small impedance match transformer to drive the transducer as shown in Figure 79.

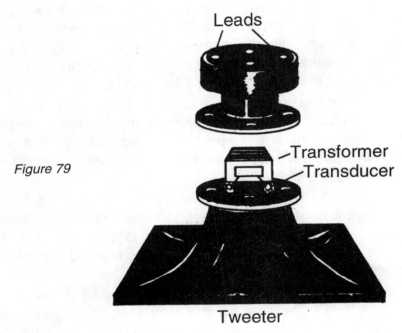

Figure 79

Where they are found

Piezoelectric transducers are found in many applications. Quartz crystals are found in computers, watches, clocks, transmitters, and many other applications. The sound transducers are found in earphones, alarms, audio equipment, etc.

Testing

Quartz crystals are tested using proper circuits. The other transducers (ceramic) are tested using oscillators to drive them.

Section II—Electronic Components

Semiconductors

Today, any practical application of electronics involves the use of semiconductors. Transistors, diodes, and their relatives form the great semiconductor family found in most any type of modern electronic equipment. With help, the electrician can understand how these components work and why it is important to start with some fundamentals of semiconductors.

Semiconductor Principles

According to modern physics all substances can be placed into one of three classes: insulator, conductor, and semiconductor. The main factor that determines in which of these classes the substance can be placed is the spacing of the energy band in its molecular makeup. Each energy band can hold two electrons at any time.

If the energy bands of a substance are filled, that substance cannot accept or donate electrons and so is an insulator. If there is one electron per band, or if the bands are so closely spaced that no gap exists between a filled band and an empty band, electrons can move through the material and it is a conductor. Now, the intermediate case: If small gaps exist between filled energy bands and vacant ones, the material acts like an insulator at low temperatures and becomes a conductor when the temperature rises. This is a semiconductor material.

Semiconductor types such as the silicon, germanium, gallium, and others have additional properties that make them ideal for use in electronics. They are elements that have four valence electrons each. Because of the valence bonds, they form a basic crystal structure as shown in Figure 80.

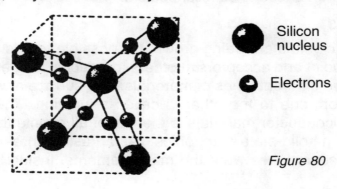

Figure 80

Any crystal is held together by the sharing of electrons between the atoms. In a conducting material the free electrons can drift through the structure under the influence of electric forces. On the other hand, in a semiconductor, availability of a path for free electrons to move along depends upon temperature. As the temperature rises, the path becomes available and high-energy electrons can move.

A crystal of semiconductor like silicon or germanium is composed of billions of atoms bonded together in an arrangement similar to the one shown in Figure 80. Crystals of pure semiconductors can be grown in laboratory conditions. They are called intrinsic materials and have no practical use. But, if we add small amounts of specific impurities to the intrinsic material, these impurities have the capability of joining right into the crystal structure at the atomic level.

There are two kinds of impurities resulting in different effects on the electric properties of the material. If a five-electron element, such as the antimony of boron and phosphorus, is added to the crystal, each of its atoms has one valence electron left over that finds no partner in the crystal structure. The net result is a surplus of electrons. Elements such as this are called donor substances and their presence provides an excess of negative charges. The semiconductor crystal is now called an n-type material.

Now, if we add a three-electron element, like aluminum, gallium, and indium, to the structure of the crystal, the net result is the presence of holes where the surrounding semiconductor atoms have electrons, but the impurity has no matching electrons to fill the valence bond. Elements with three electrons in the valence shell are called acceptors and form p-type materials when added to semiconductors.

Junctions

When crystals are being grown it is possible to add donor materials at one point and acceptors at others, producing single crystals composed of two different types of materials. The surface at which the type changes from one to the other is called a junction. The junction between semiconductor materials is the key of all events that made possible modern solid-state electronics. (As an aside, in early electronics circuits, vacuum tubes were the main elements. Inside these circuits, all events occurred in the vacuums. Today, the main elements of any circuit are semiconductors—solid materials—thus, the name "solid-state" electronics.)

Combining junctions and different semiconductor materials, engineers have created a large number of components that are found in every kind of electronic appliance. Let's take a look at the main types of semiconductor components, how they work, and where they are found.

Diodes

The first important electronic component of the semiconductor family is the diode. To understand how a semiconductor diode functions, look at what happens in a semiconductor junction such as the one described previously.

At a junction the surplus electrons of the n-type material diffuse across the junction filling the holes of the p-type material. This process is called combining and forms valence bonds that can no longer diffuse throughout the crystal. This means that in a small area, a depletion region is formed that is free of holes and electrons.

If, in a semiconductor junction, we apply a positive voltage to the n-type part of the junction, it tends to draw electrons out from the n-type material. At the same time, the electrons in the negative side of the power supply fill the holes in the p-type material. The net result is that the depletion region expands to the entire structure of the material. This situation prevents current flow. The diode is now reverse-biased.

If we reversed the applied voltage, the electrons would be forced away from the power supply toward the junction, and, at the same time, holes from the power supply forced to the junction. The electrical force involved in this process compresses the depletion region until it disappears. At this moment, the barrier is broken and the current can flow through the device. With germanium devices this occurs at about 0.2 V and with silicon devices at about 0.6 V. This way the junction is forward-biased.

The device formed by a single p-n junction is called a diode. Figure 81 shows what happens in the two described bias conditions.

Electronics for the Electrician

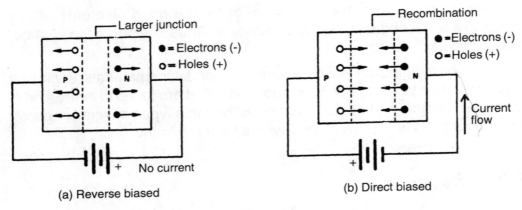

(a) Reverse biased (b) Direct biased

Figure 81

The practical semiconductor diode is formed by two pieces of semiconductor materials—n- and p-type germanium or silicon—placed inside an enclosure. The size of the materials is basically determined by the amount of current that they can conduct when forward-biased. Semiconductor diodes operate as one-way roads for the current and are found in a large number of electronic circuits.

Symbol and Types

The diodes are manufactured in different sizes and types according to their applications. Figure 82 shows common diodes and the symbol used to represent this component.

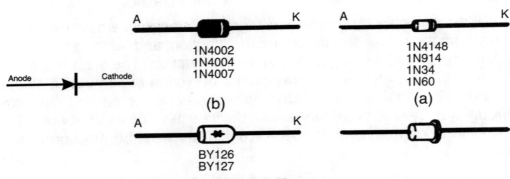

Figure 82

Section II—Electronic Components

In (a) are germanium and silicon small-signal diodes that are used to handle low currents. In (b) are some rectifier diodes used in circuits where large currents are found.

Diodes are polarized components. Their position in a circuit is important. The identification of the poles is made using several resources. One method is placing the symbol and the component to correspond with the position of the poles and other is to use a strip or ring at the side of the cathode terminal.

Specifications

A. Reverse voltage: Since a diode represents an open circuit when reverse-biased, across the diode appears all the voltage of a circuit. The maximum voltage that a diode can be submitted to without being burned under these conditions is an important parameter to be considered in projects. This voltage is also abbreviated by Vrrm or Vr in data sheets.

B. Direct current: This is the largest current that can flow through the diode when it is forward-biased. It is also abbreviated by If.

C. Part number: The most common in real components is the manufacturer specification by a part number. In most parts of America, specifications of diodes are identified by the group "1N" followed by a number. So, types such as 1N914, 1N4148, and 1N4002 are common. In European specifications diodes begin with an "A" (for germanium) or "B" (for silicon) followed by a letter. For example, "A" for general-purpose diodes, "Y" for rectifiers, etc. A European part number might look like these: AA115, BA315, BY127, etc. Other manufacturers use proper codes to identify their diodes. Some examples are MR751 and P600D.

Where they are found

The diodes are used basically in functions such as detectors or signal clippers in audio and RF circuits or as rectifiers in power supplies.

As detectors, the electrician will find small diodes (silicon or germanium) in radio receivers, TV, remote-control receivers, computers, interfaces, wireless telephones, etc. As signal clippers, the diodes are used to convert an AC into a DC as shown by the diagram in Figure 83.

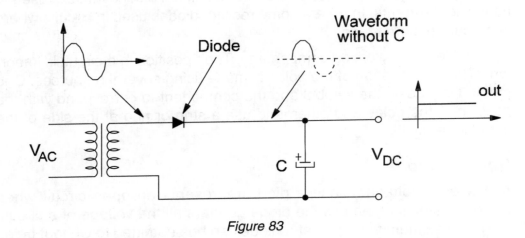

Figure 83

In this circuit, the diode conducts only the positive half-cycles of the AC voltage found in the secondary winding of the transformer. Filtering these positive pulses with a capacitor, a DC voltage can be placed at the output. Power supplies such as this are used to power small DC appliances like calculators, CD players, and transistor radios from the AC power line. Figure 84 shows an AC/DC adapter using an inside circuit like this.

Figure 84

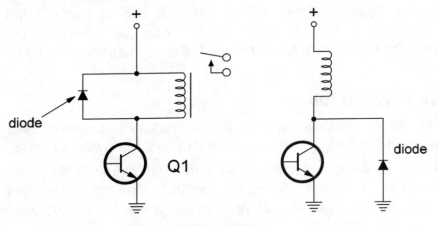

Figure 85

Section II—Electronic Components

A third case to be included here, and one that is very important to electricians, is the use of a diode to protect semiconductors from voltage spikes generated when inductive loads are triggered. When a relay is open or a solenoid is turned off, it generates a voltage spike strong enough to damage the device used in the control circuit. The energy produced in this process can be absorbed by a diode wired in parallel with the load or with the control device as shown in Figure 85.

Testing

A diode can be tested using a common multimeter or any continuity tester. Adjust the multimeter to a medium resistance scale and place the probes in the diode's terminals. When biased forward the diode must present a low resistance (between 100 and 5000 ohms) and when reverse-biased a high resistance is read (above 1,000,000 ohms). A diode presenting low resistance when reversed and forward-biased is shorted, and the one that presents a high resistance of the two measurements is open. An intermediate read when reverse-biased (between 100,000 and 500,000 ohms) indicates a diode presenting losses.

Zener Diodes

When a diode is reverse-biased a limit value to the applied voltage exists. Beyond this limit, called breakdown voltage, the diode becomes a conductor and is destroyed as shown by the graphics in Figure 86. As shown by the curve, the diode, if not destroyed, tends to keep the voltage constant across its terminal in a range of currents.

Zener diodes are special diodes that are processed to have a nondestructive breakdown at a specified value of the reverse voltage. They are also called avalanche breakdown diodes or breakdown diodes.

The action of a zener diode can be used in many electronic applications. In fact, if you apply a reverse voltage to a zener diode its reverse resistance changes in a way to maintain constant voltage across it. This action can be used in voltage regulation, to clip signals, and many other applications.

Zener diodes are found in many types and sizes in electronic equipment. For the electrician, it is important to know how a common diode can

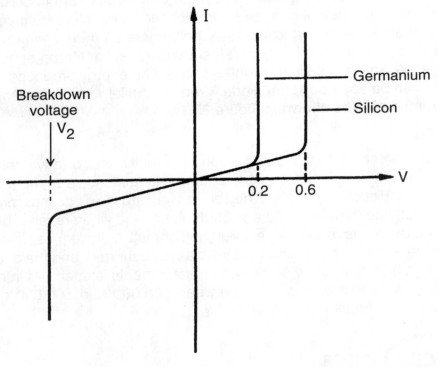

Figure 86

differentiate from a zener diode. Although they look the same, the way they are positioned in a circuit, the identification number, the symbol in the schematics, or any other specific information may differ.

Symbol and View

Figure 87 shows the symbol and a view of a common zener diode as the electrician will find them in electronic circuits.

Since zener diodes are polarized components, there must exist some identification for the poles in the component. It is usual to place a ring or strip at the cathode side of the component.

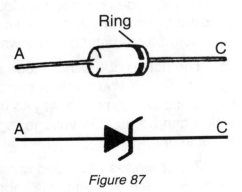

Figure 87

Section II—Electronic Components

The size of the zener diode is determined by the amount of current it can control or the range of currents in which the diode can keep the voltage constant.

Specifications

The zener diode manufacturers used to indicate their products by a part number. So, a handbook with the main types or another kind listing of their characteristics is fundamental in order to know what they are. When replacing a zener diode without having the part number it is important to know two specifications:

1. Zener Voltage: the breakdown voltage or the voltage the device is designed to keep constant when reverse-biased. Zener diodes are made with voltages in the range between 1 and 2 volts to more than 100 volts for typical types found in domestic and industrial appliances.
2. Dissipation Power: the maximum amount of current flowing across a zener diode during operation. The zener voltage gives the dissipation power or the amount of electric power that the device can convert in heat. The larger the size of the device, the larger the dissipation power. Typical values are between 400 mW and 10 W.

Where they are found

Figure 88 shows some circuits where zener diodes are found. In (a) is a simple voltage divider in a simple power supply where the zener diode maintains a constant voltage in the load plugged into the output. A more

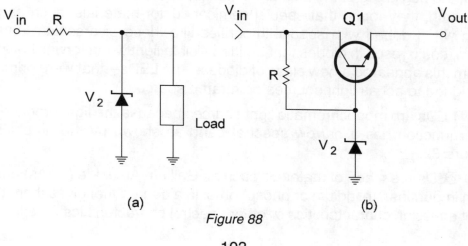

(a)   (b)

Figure 88

complex power supply is shown in (b) where the transistor controls the amount of voltage applied to the load. This voltage is kept constant by the action of the zener diode placed at its base. How the transistor works in a circuit like this is discussed when explaining how transistors work.

Other applications of zener diodes include voltage reference voltage in instruments and signal clipping circuits.

Testing

The simplest way to test a zener diode uses a common multimeter on the lowest resistance scale. The zener diode is tested as a common diode; when forward-biased it must present a low resistance (100 to 10,000 ohms), and when reverse-biased it must present a high resistance (above 1,000,000 ohms). This test doesn't show anything about whether the zener voltage is altered or not. It is important to remember that this test is valid only if the zener voltage of the tested device is lower than the voltage used by the multimeter in the probe.

LEDs

Light Emitting Diodes or LEDs are special diodes that produce light when forward-biased. The current flowing to their PN junction causes the emission of visible light or infrared radiation. Researchers discovered that any semiconductor diode produces infrared light when conducting a current, the beginning of the LED's principle of operation. Advancing with the research, they found that special semiconductor materials, such as gallium when doped with special impurities like those of arsenium and indium, could result in diodes that emitted visible light when forward-biased. From this appeared a new class of diodes—the LEDs—that were primarily designed to act as light sources or infrared sources.

LEDs are monochromatic light sources because the light produced in their junction has a narrow spectral band as shown by the graphics in Figure 89.

LEDs are made of materials such as Gallium Arsenide (GaAs) doped with impurities like indium or phosphorus, in a degree that gives them their light emission characteristics or other spectral characteristics. The impuri-

Section 11—Electronic Components

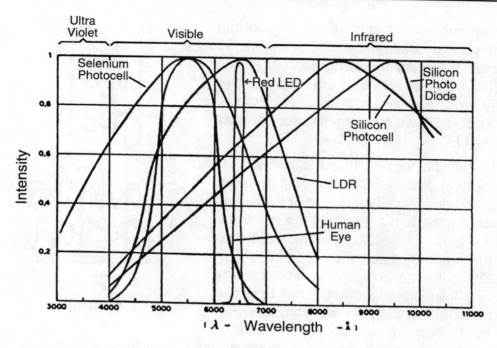

Figure 89

ties also determine the minimum voltage for forward-biasing an LED that makes it conductive. Red LEDs become conductive with 1.6 V; yellow LEDs need about 1.8 V, and blue and green LEDs need more than 2.0 volts.

Symbol and View

Figure 90 shows the symbol used to represent LEDs in schematic diagrams and in the same figure is the main type of LED.

The different types and sizes depend on the color of the light and also the application. IR (Infrared) LEDs are used in remote controls and as sensors. LEDs can also be placed to form an eight in a

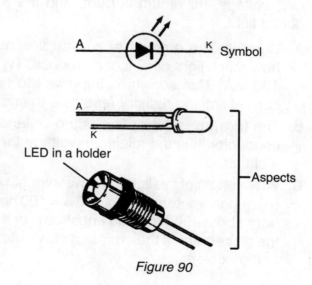

Figure 90

seven-segment display. Each segment in this kind of display is an LED. Figure 91 shows how they are connected. If all the LEDs are wired with the cathodes together they are called common-cathode displays. Figure 91 shows a seven-segment display using LEDs.

It is also possible to make a white LED by putting three LEDs on a chip with the basic colors (Red-Blue-Green or RGB) and powering them in a way that the combination of their light results in white. According to the amount of current in each of these LEDs, it is also possible to produce any color of the visible spectrum. The LEDs, like the diodes, are polarized components. Normally they are encapsulated in clear or colored plastic with a small, flat space or notch near the cathode lead.

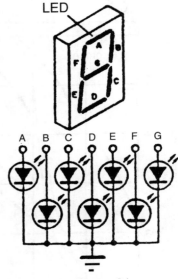

Figure 91

Specifications

LEDs are specified by a part number provided by the manufacturer. But, catalogs also indicate other characteristics including the size, forward-bias voltage, maximum current, and the wavelength or color of the produced light.

A. Current: The amount of current flowing through an LED determines how much light or IR it can produce. Typical values are between 5 and 100 mA. This current is important to help the designer calculate the current limiting resistor (see how to use an LED).

B. The forward voltage: This is the voltage at which the LED becomes a conductor. It is the minimum voltage that can be applied to the LED in a circuit.

C. Wavelength of the light: The wavelength can be expressed in angstroms (Å) or nanometers (nm). 1000 A = 100 nm. The visible spectrum extends from about 4000 to 7000 angstroms or 400 to 700 nm. Figure 92 shows the placement of the emission curve of some common LEDs according to their colors.

Section II—Electronic Components

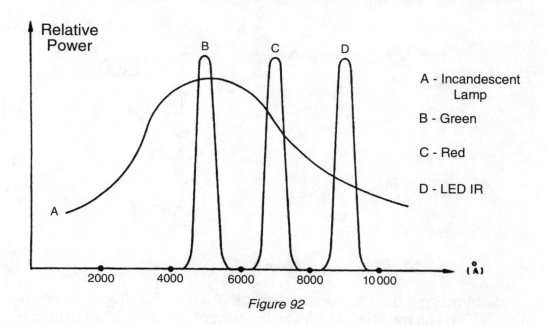

Figure 92

Where they are found and used

The main applications of LEDs in electronic equipment are as indicators. The electrician will find LEDs in panels of electronic, and even electric, devices indicating a function or displaying a number or even a figure as in the seven-segment types. The seven-segment displays also indicate the result of operations in calculators, the hours and minutes in clocks, time, and other information in microwave ovens, etc.

Infrared LEDs are found in remote controls of a variety of equipment like TVs, VCRs, automatic doors, car alarms, etc.

When using LEDs it is important not to exceed the maximum allowable forward current. Current is best limited by a resistor placed in series with the LED and the power supply as shown in Figure 93.

The resistances of the series resistor can be calculated by subtracting the LED's forward voltage from the power supply voltage and dividing the result by the desired current in amperes. For example, if we want to power a red LED at 20 mA from a 6 V power supply the math is as follows:

Electronics for the Electrician

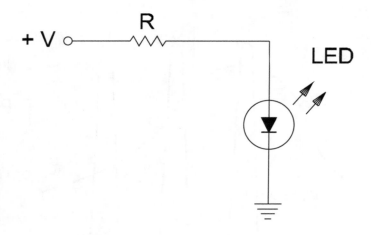

Figure 93

Subtracting 1.6 V (the forward voltage of a red LED) from 6 V results in 4.4 V. Dividing the result (4.4 V) by the current 20 mA (0.02 A) results in 220 ohms.

Testing

Never test an LED by plugging it into a power supply without a series current-limiting resistor. It can be burned. The best way to test is by plugging it into a 3 V supply (2 AA cells) in series with a 100 ohms to 270 ohms resistor as shown in Figure 94. The LED will glow if it is good. To check whether an infrared LED is producing radiation, some kind of detector must be used.

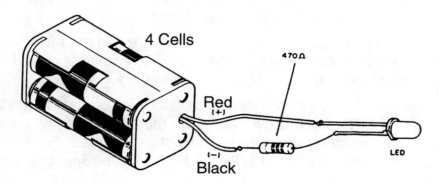

Figure 94

Special Diodes

The electric properties of a PN junction can be used in devices other than the LEDs. Many special diodes can be included in a list:

A. Photodiodes: Exposing the PN junction of a diode to light and reverse-biasing it, the current flow will depend on the amount of light falling on the junction region. Charge carriers are liberated by the light, increasing the reverse current. The semiconductor material with a PN junction is then manufactured in transparent enclosures and used as light or IR sensors. These devices are called photodiodes.

B. Tunnel diodes: Tunnel diodes are special devices presenting the characteristic of a curve in a negative resistance region. This characteristic makes the device ideal to produce very high-frequency signals. The tunnel diode is used as UHF (Ultrahigh Frequency) and SHF (Superhigh Frequency) oscillators and detectors in many applications like radar detectors.

C. Varicaps or varactors: When reverse-biased, a diode can be compared with a capacitor. The junction region is the dielectric and the PN materials with the plates as shown in Figure 95.

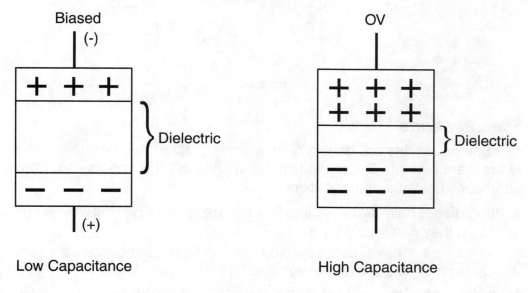

Figure 95

Electronics for the Electrician

Depending on the applied voltage, the size of the junction region can be altered. If a low voltage is applied, the charges keep the junction region thinner. But, by increasing the voltage, the charges make the junction region fatter. This means that by applying a voltage in a reverse-biased diode we can control its capacitance.

Special diodes with large PN junctions (to get more capacitance) are fabricated to be used as variable capacitors controlled by voltages. They are called variable capacitance diodes or Varicaps. These capacitors are found in tuning circuits of TVs and radios.

Symbols and Types

Figure 96 shows the symbols and types of some special diodes. Remember that these diodes are also polarized components and some mark is used to the identification of their poles.

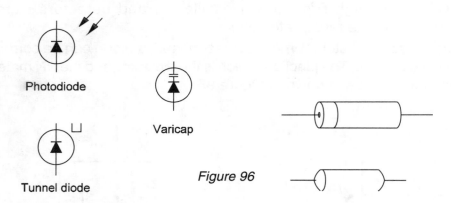

Figure 96

Specifications

Special diodes are generally indicated by a part number determined by the manufacturer. Depending on the application, it is important to know some specifications of these diodes.

- Photodiodes have, as a specification, the response curve (the wavelength of light they can "see").
- Tunnel diodes have, as a specification, the highest frequency they can operate and also the tunnel voltage.
- Varicaps can be specified by the capacitance range they can present between zero and a nominal voltage.

Testing

The test of many special diodes cannot be done using simple procedures. We can detect if a photodiode, tunnel diode, or varicap diode is shorted or open, but not determine any other important characteristics of these components, such as whether they are operating, for example.

Bipolar Transistors

Bipolar transistors are the most important of the semiconductor devices. They are formed by three pieces of semiconductor materials, forming two junctions in the same crystal with a structure such as the one shown in Figure 97.

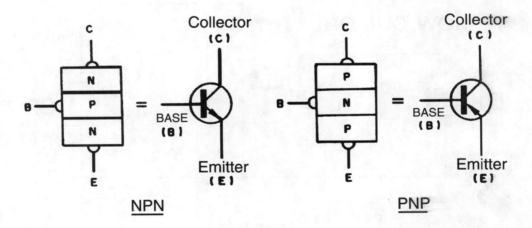

Figure 97

Although the structure can be compared with two diodes placed back-to-back, the fact that they are in the same crystal means the events occurring at one junction can influence those at the other. The possibility of control makes the transistor one of the most important of all the electronic components.

According to the type of semiconductor, the structures result in two types of transistors—NPN and PNP.

To each piece of semiconductor material, terminals or leads are placed corresponding to the three regions: base (B), emitter (E), and collector (C). The transistor forms a structure with special electric properties; a current flowing between emitter and collector can be controlled by a current flowing through the base. When connecting the positive pole of a battery to the emitter and the negative to the collector of an NPN transistor, because the two diodes are reverse-biased, no current can flow.

But, if the base-emitter junction is now forward-biased in a manner that a small current can flow, this current will cause a high current to flow between collector and emitter as shown in Figure 98.

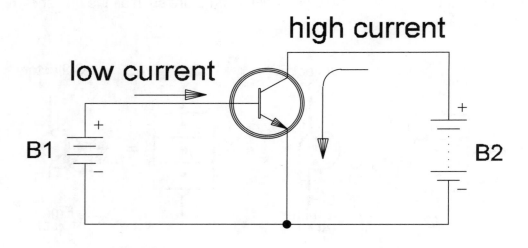

Figure 98

In other words, a small current flowing between the base and the emitter of a transistor can control a large current between collector and emitter. If a 1 mA current through the base causes a collector current of 100 mA, the transistor has a "gain" or "amplification factor" (also called Beta) of 100. This means that the transistor can be used to amplify signals or to act as an electric switch.

Based on the applications, we can find transistors of different sizes and formats with gains ranging from 5 to 10,000. Today there are millions

Section II—Electronic Components

of types of transistors identified by a manufacturer part number. Only with heavy, large "databooks" or "transistor manuals," could the characteristics of all the main transistors be found.

In applications involving transistors, the use of about 100 or 200 main types with characteristics that are near most of other existing transistors is normal. These transistors can be found easily at a dealer, and in emergency cases, can be replaced with any other.

Symbol and Type

Figure 99 shows the symbols used to represent the two types of bipolar transistors and the common types of these devices. Notice the arrow in the emitter points outside in the NPN type and inside in the PNP type.

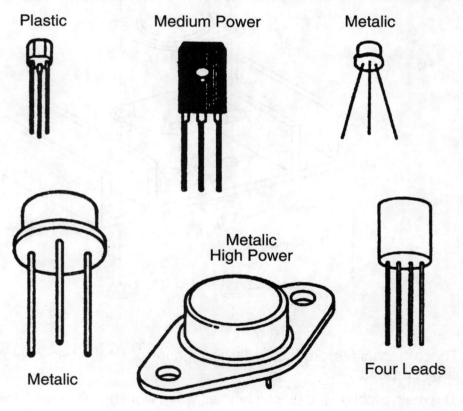

Figure 99

The size of a transistor is determined by the amount of current it can control. Transistors called general-purpose transistors or low-power transistors are small and encapsulated in plastic as shown in (a). Medium-power transistors are shown in (b). These transistors have a hole or another way to place them in heat sinks. In (e) is a high-power transistor that is mounted in a large heat sink. In (f) is a four-lead transistor. The fourth lead is the case acting as screen.

Figure 100 shows a high-power transistor like the one shown in (c), mounted on a heat sink.

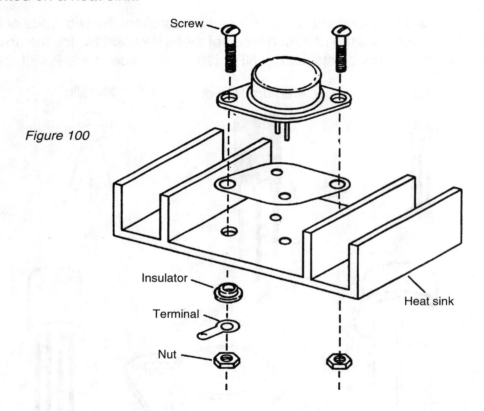

Figure 100

These transistors produce a large amount of heat when operating with large currents.

The main problem the electrician has when working with transistors is the identification of the terminals or leads (emitter, collector, and base). Many special codes and marks are used for this task such as flat spaces, shafts, marks, color points, and others. With the specific information about one transistor, it is possible to know which terminal does what. Otherwise, using the multimeter is also a possibility to make identification by measuring the resistance between the terminals.

Specifications

Besides the size and type (NPN or PNP), the bipolar transistor can also be classified according to the application. The basic groups according to use are:

A. Low-frequency or audio: Low-frequency or audio transistors are components intended for use in applications where the signals are in the audio band or below some hundreds of kilohertz as in audio amplifiers, power supplies, etc. They can handle low-frequency signals such as the ones found in small amplifiers and the audio stage of radios.

B. High-frequency or RF (Radio Frequency): High-frequency transistors are devices specially intended to operate with signals ranging from some megahertz to gigahertz. They are found in radios, TVs, remote controls, wireless telephones, etc., in the stages where the high-frequency signals are present. They also can be found in sizes according the power to be handled.

C. Switching: Transistors that perform this action are used as switches in high-speed circuits. They are used to turn devices on and off in many applications. They can also appear in low-, medium-, or high-power versions. They are found in inverters, logic circuits, digital instruments, digital controls, etc.

Other specifications

The manufacturers indicate the transistor by a part number. With the part number, an electrician can find all the electric characteristics of the device needed to know what it does in a circuit. This is especially important when replacing a transistor because it is possible to find an equivalent or replacement type. The equivalent is a transistor with another part number, either from the same or another manufacturer, but with all the characteristics as the original to replace it in an application.

It is very important to remember that a transistor that is equivalent to another one in application can't be equivalent in another application or circuit!

The main electrical specifications of a transistor are:

A. Maximum voltages: Vce(max) is the maximum voltage that can be applied between collector and emitter. Vceo(max) the same, but with open base. Vcb(max) is the maximum voltage that can be applied between collector and base. Vcbo(max) the same, but with open emitter. Veb(max) is the maximum voltage that can be applied between base and emitter. Vebo(max) the same, but with open collector. In some case the absolute values can be specified as Vc, Vb and Ve.

B. Maximum currents: The most important current in a transistor is the collector current. It can be specified as Ic or If (max).

C. Maximum power: When conducting a current, the transistor generates heat that can be transferred to the air. The maximum amount of heat produced by a transistor is given by its maximum dissipation power. These specifications are given in watts. Small transistors dissipate power in the range between 50 and 500 mW. High-power transistors have power specification above 100 W. The parameter is indicated by P or Pd in the datasheets of transistors.

D. Gain: A small current forced to flow by the base element causes a larger current to flow between emitter and collector. The number of times the collector current is larger than the base current is the gain of the transistor or beta factor (β). If a 1 mA current in the base of a transistor causes a 100 mA current between emitter and collector the gain of this transistor is 100. Normally, the transistor has a specification range of gain or minimum gain. For example, the transistor 2N3905 has a gain range between 50 and 150. The gain also can be indicated by a hybrid parameter named hFE, which is equivalent to the beta factor.

E. Transition frequency: The gain of a transistor falls when working with high-speed signals. But, there is a frequency where the gain falls to unit. This means that in this frequency, a variation of 1 mA in base current causes a variation of 1 mA in the collector current, and the transistor cannot operate as an amplifier. The output current is the same as the input current. This frequency is called transition frequency or fT.

Working with transistors

The electrician who is not very experienced with electronics sometimes can be confused when working with transistors. Millions of types and too many characteristics to observe when trying to replace one are the main factor of this embarrassing situation. It is even more aggravating when one considers that they have three terminals, not just two as the others.

How are the transistors used? How do you test a transistor? How do you find an equivalent? How do you know how they work? How do you know which is the base, collector, and emitter leads?

As the transistor is one of the most important of the electronic components in any electronic device, more information will be provided on this component later.

How the transistors are used

Figure 101 shows how a transistor is used in the two basic operation modes (linear and saturated). In (a) the load is placed between the power supply and the collector of an NPN transistor. The resistor in the base determines the amount of base current. This resistor is called the biasing resistor. When closing S1 the base current causes the current to flow through the load.

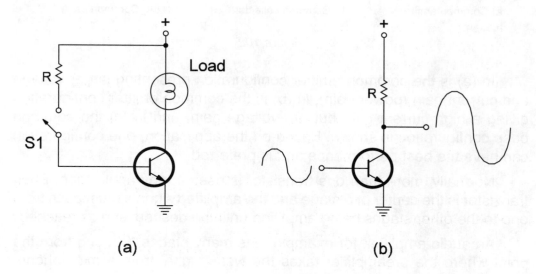

Figure 101

If as shown in (b), a variable current as the correspondent to an audio signal is applied to the base, the small variations of this current will produce large variations in the collector current.

The way the transistors are biased and how the signals to be amplified can be applied and taken off a circuit determine three basic configurations of the transistors. The three ways or basic configurations are shown in Figure 102.

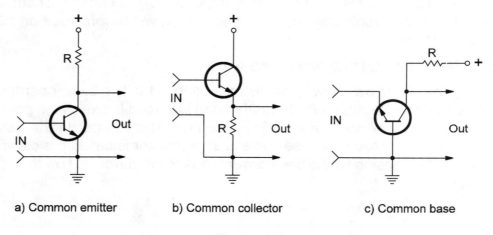

a) Common emitter   b) Common collector   c) Common base

Figure 102

In (a) is the common emitter configuration presenting larger voltage and current gain (power gain), in (b) is the common emitter configuration giving a high current gain but low-voltage gain, and in (c) the common base configuration is shown. Based on the application, one configuration can have the best performance and is preferred.

Normally, more than one transistor is used in any application. Each transistor is the center of a stage and the amplified signals must pass from one to the other stages being amplified until the desired level is reached.

An audio amplifier, for example, has many stages. Starting from the point where the preamplifier takes the weak signal from a microphone, passing it to intermediate amplification stages, and ending in a power output stage, it can source enough power to drive a loudspeaker reproducing sound with a desired volume. Figure 103 shows a typical circuit of an audio amplifier using two transistors.

Section 11—Electronic Components

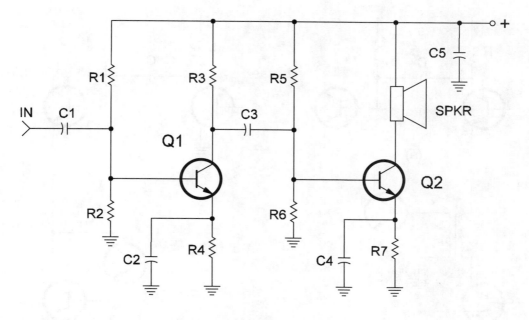

Figure 103

The way the transistor transfers the signals from one stage to another is called coupling, and there are several coupling methods as shown in Figure 104.

In (a) is an RC coupling. The signal is transferred from one transistor to another by a capacitor. In (b) is an LC coupling. The capacitor is also used to transfer the signal from one stage to another. In (c) is a transformer coupling, and in (d) is the direct coupling. A special method of direct coupling, called a Darlington coupling (not shown), is when two NPN or PNP transistors are used.

Testing and identification

The electrician who is just starting with electronics may find some appliances using transistors that are in trouble. How should one proceed in this case? Transistors are found practically in all equipment, forming the central element of many stages. Each stage can have as core one or more transistors in one of the configurations we have seen before. The main test that the electrician can do is measuring the voltages in the leads of the transistor using a multimeter.

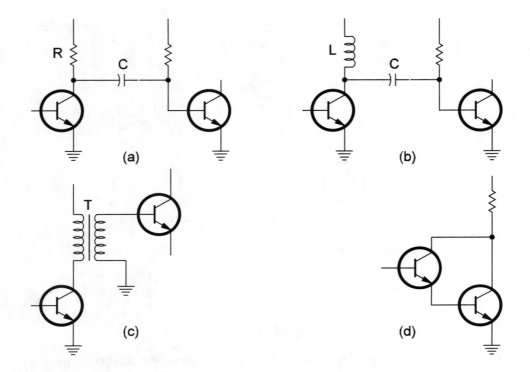

Figure 104

As a general rule, the collector voltage is higher than the emitter voltage in an NPN transistor and the inverse in a PNP transistor as shown in Figure 105.

The base voltage is 0.6 V to 0.7 V above the emitter voltage in an NPN silicon transistor and the same below the emitter voltage in a PNP silicon transistor. In old germanium transistors the difference is 0.2 V. If the voltages are different, the biasing resistor or other elements surrounding the transistors can have problems.

The static test of a transistor or the test out from a circuit, is done by measuring the resistance between the elements (EBC) using a multimeter. It is a "junction test" where it is verified whether the two junctions (base-emitter and base-collector) are okay.

The basic test consists of "seeing" if the two "virtual" diodes corresponding to the two junctions are good. Simply adjust a multimeter to a

Section 11—Electronic Components

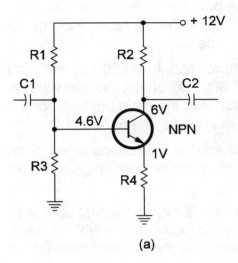
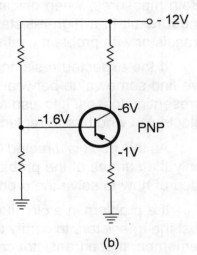

Figure 105

low-resistance scale and make six measurements in on the transistor to test it. Figure 106 shows how to do these measurements.

When placing the probes between base/collector and base/emitter one measure must be of a high resistance and the other a low resistance.

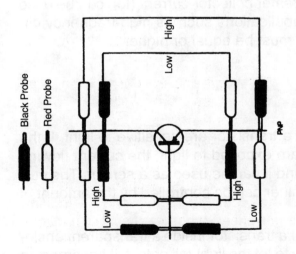
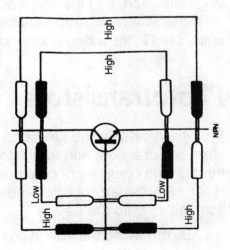

Figure 106

121

Both measures, when placing the probes between collector and emitter, must result in a high-resistance reading. Any different result indicates a transistor with problems, either shorted or open.

If the expected resistance is high (more than 1,000,000 ohms), and we find some value between 20,000 and 500,000 ohms, the transistor is presenting losses. It is also a bad transistor. But, the main problem for the electrician who has to replace a transistor is how to identify it.

As this book is directed toward beginners, the indications given touch only the surface of the problem, although they are enough to give you an idea of how to solve the problem.

If a problem in a circuit using a transistor is found, the first step is to test the transistors to certify that they are the cause of the trouble. Always remember that a transistor can burn if associated components (capacitors and resistors) fail. A capacitor, in short, can cause high currents across a circuit burning a transistor. So, before replacing a burned transistor, test all the components near it as well.

For the replacement, if the original cannot be found, look in a handbook for equivalencies, replacement types, or the characteristics. As a practical rule, a transistor for the same function (small signal for instance) and the same type (NPN or PNP) with the same or greater gain, the same or greater Vce, and the same or greater collector current (Ic), can be used as replacement. In some critical applications such as high-frequency circuits, the fT (transition frequency) must be equal or higher.

Phototransistors

As in diodes, the junctions of a transistor are sensitive to light. If the junctions of a common transistor are exposed to light, the current flowing through the component changes and it can be used as a sensor. The current flow between collector and emitter can be controlled by the amount of light falling onto the device.

By mounting the silicon chip of a transistor inside a transparent enclosure or an enclosure with a window to let the light fall onto it, it becomes an important electronic device—the phototransistor.

Section II—Electronic Components

Phototransistors are used the same way as the photodiodes—as sensors in many electronic applications. One important aspect of phototransistors when compared with LDRs is that they can be used to detect very fast changes in the light intensity.

Symbol and Types

Figure 107 shows the symbol adopted to represent the phototransistor and a few types. The phototransistor can "see" light invisible to the human eyes. They can be used to detect both IR (infrared) and, in some cases, UV (ultraviolet).

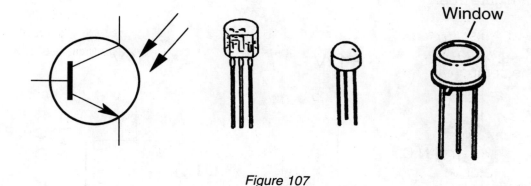

Figure 107

Specifications

The manufacturers mark their phototransistors with a part number. Characteristics such as wavelength response, gain, collector current, maximum operating voltage, and others must be found from using the part number to consult the manufacturer's documentation.

A common series of phototransistors are the one made by Texas Instruments. The first letters (TIL) indicate that it is an optoelectronic device and the next number indicates a special type. Examples of these part numbers are TIL81 and TIL411. The same initial letters are used to indicate other components working with light, such as LEDs and optocouplers.

Where and how they are used

The electrician will find phototransistors in many applications related to professional activities. In fact, alarms, motion detectors, remote controls, and many other applications use the phototransistor as a sensor.

Because the phototransistor is much faster than the LDRs, they have different uses. The electrician will find the LDR in light detectors where the changes or amount of light are slow as in emergency lights and automatic nightlights. In the applications where fast changes of light must be detected or where information must be sent through a light beam (visible or not) the photodiode and the phototransistor are preferred.

In a typical application the phototransistor is used as a diode, as shown in Figure 108.

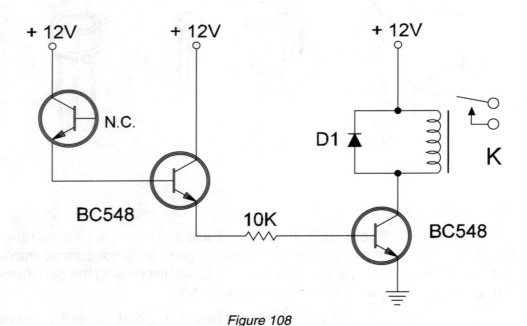

Figure 108

In the circuit the base terminal is let free and the component acts as a photodiode. The current flowing between emitter and collector depends on the light picked up by the device. In many cases, although the component is a transistor, the package doesn't present the base terminal.

Section II—Electronic Components

In order to increase the amount of light picked up by a phototransistor a lens can be placed in front of the transistor. It is important to place the transistor in its focus as shown in Figure 109.

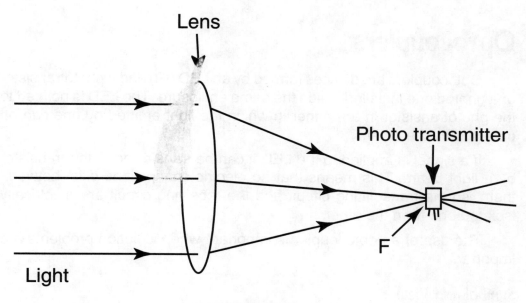

Figure 109

A typical application of a phototransistor and an infrared LED is shown in Figure 110.

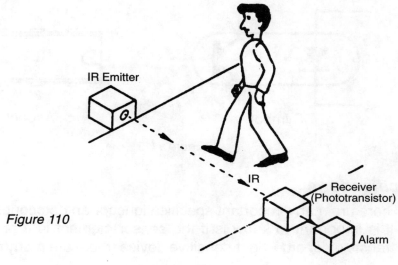

Figure 110

Electronics for the Electrician

The infrared LED sends a light beam that is focused on the phototransistor. If the light beam is interrupted (cut by an intruder) the circuit detects the cut and triggers an alarm.

Optocouplers

Optocouplers are devices formed by an LED (IR) and a phototransistor or a photodiode installed inside the same enclosure. The LED is pointed to the phototransistor in a manner in which the light emitted by one can be detected by the other.

If a signal is applied to the LED it can be transferred to the transistor by a light beam. This means that no electric connections exist between them, so the transmitting circuit and the receiving circuit are electrically isolated one from the other.

The use of an optocoupler is important when isolation problems are important.

Symbol and View

Figure 111 shows a view and the symbol used to represent an optocoupler.

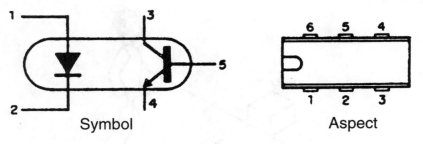

Symbol   Aspect

Figure 111

Specifications

There are many important specifications for an optocoupler. The first thing it is important to know is if the sensor element is a photodiode, a phototransistor, or other light-sensitive device (there are many others). The part number is important, but gives no information on this topic. The electrician will need a handbook with the characteristics or documentation from the manufacturer.

The main characteristics and their meanings are:

A. Isolation voltage: This is the maximum applied voltage between the LED circuit and the sensor. Typical values range from 3000 to 7000 volts.
B. LED characteristics: These are the maximum forward current and voltage.
C. Sensor characteristics: This is the maximum voltage between collector and emitter if a transistor is used.

Where they are used

The electrician can find optocouplers in applications where a signal coming from a circuit or the power supply line must be isolated from other circuits. For example, a circuit that can be triggered from the AC power line voltage, but must not be connected directly to it by security requirements. Another application is interfacing PCs and microprocessors with circuits powered from the AC power line. Using an optocoupler, the delicate circuits of the PC are isolated from the AC power line. If something goes wrong with the circuit, the PC is protected against damage.

Testing

The optocouplers can be tested by independent verification of the LED and the phototransistor. Another way is to excite the LED from a power source (doesn't forget the limiting resistor) and see if the resistance between collector and emitter of the internal transistor falls. A multimeter on the highest resistance scale can be used in this test.

Darlington Transistors

Two transistors wired as shown in Figure 112 form a Darlington stage or a Darlington Amplifier. This stage acts as a super transistor or a single transistor with gain corresponding to the total product of the gains of the individual transistors.

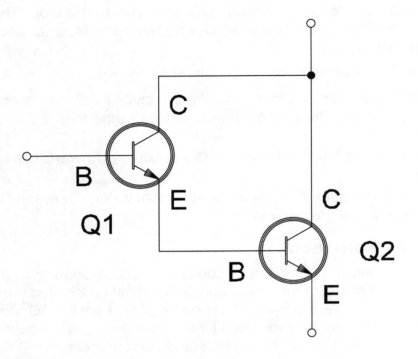

Figure 112

For example, if two transistors with gains equal to 100 ($\beta = 100$) are placed together to form a Darlington pair the final gain will be 100 x 100 or 10,000! They function as a single transistor with 10,000 of gain.

A Darlington transistor can be manufactured by placing two transistors in the same silicon chip, packing them inside a common enclosure forming a super transistor. Darlington transistors are useful where high gain is necessary, such as in solenoid, motor, relay, or lamp drivers as well as high-power audio amplifiers.

Symbols

The symbols for Darlington transistors are shown in Figure 113. As a common bipolar transistor, they are found in the NPN and PNP versions. In (a) is a low-power general purpose Darlington transistor and in (b) a high-power type that must be mounted in a heat sink.

Section 11—Electronic Components

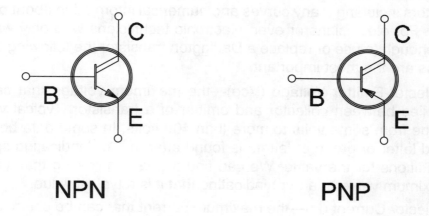

Figure 113

By the simple look of the component it isn't possible to know if it is a common transistor or a Darlington transistor, or any other semiconductor because they all use the same bodies. The identification is made by the type or part number that is normally a figure on the component's body.

Specifications

If you have a handbook of transistors or the specific data sheet (or diagram) for the equipment, it is possible to know if the component is a Darlington transistor.

A common series of Darlington transistors found in many electronic appliances and even in electric applications is the TIP series of Darlington transistors first manufactured by Texas Instruments and now by many other manufacturers. This series is formed by a number of Darlington NPN and PNP transistors designed to control currents from 1 A to more than 10 A and voltages up to 100 V and possibly even higher.

Some Darlington power NPN transistors in this series are TIP110, TIP111, and TIP112 for 2 A of collector currents. The complementary PNP types are the TIP115, TIP116, and TIP117.

Important: Not all transistors beginning with TIP are Darlington transistors!

The data sheets or data books normally have a lot of information about transistors including many curves and numerical information about performance. For electricians or even electronic technicians who only want to know enough to use or replace a Darlington transistor the following specifications are the most important.

A. Collector-Emitter Voltage (Vce)—the maximum voltage that can be applied between collector and emitter of a transistor. Typical values range from some volts to more than 400 volts. In some data books a third letter or group of letters is found after the "ce" indicating special conditions for the value. We can find a (max) indicating that it is the maximum value or a (typ) indicating that it is a typical value.

B. Collector Current (Ic)—the maximum current that can be controlled by the device. This current flows between collector and emitter when the transistor is in the conducting state and varies from some milliamperes to more than 20 amperes to even more heavy-duty types. The additional indication of (max) can also be found with the values.

C. Gain (hFE)—Gain for common Darlington transistors can vary from 1000 to more than 10,000. For what the gain means, refer to specifications in for bipolar transistors.

D. Dissipation Power (Pd)—The transistor generates heat when it conducts a current, which is transferred to the area surrounding the transistor. The Dissipation Power in watts gives the maximum amount of heat that can transfer to the surrounding area. Small-signal transistors can have dissipation in the range of milliwatts, but high-power types can dissipate powers up to 250 watts and more.

E. Transition Frequency (fT)—Darlington transistors are slow devices and can't operate in frequencies above some hundred kilohertz. The maximum theoretic operation frequency is given by the transition frequency of fT (the gain falls to 1) and is typically near 100 kHz for most common types.

Where they are found

Many common appliances use Darlington transistors to drive high current loads including solenoids, electromagnets, motors, and lamps. This is shown in the example circuit in Figure 114.

Section II—Electronic Components

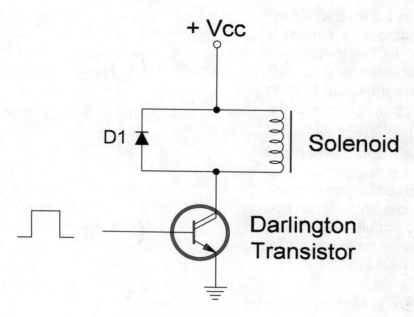

Figure 114

The load is connected between the power supply and the transistor's collector. When the transistor is biased it conducts the current energizing the load. In this configuration the transistor acts as a switch turning on and off the load according to the base current.

You can find transistors like these driving solenoids in electric appliances, electric doors, and in electronic equipment such as VCRs and DVD players. Darlington transistors can also be used as low-frequency amplifiers in other applications (Figure 115).

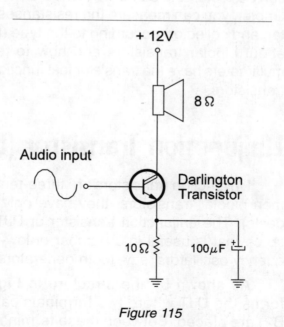

Figure 115

Electronics for the Electrician

The audio signal (corresponding to a sound) is applied to the transistor's base where it is amplified. Then the transistor triggers a loudspeaker where the signal is reproduced.

Two complementary Darlington transistors (Figure 116) are found in many hi-fi audio amplifiers. This kind of circuit can drive a loudspeaker with powers up to 250 watts.

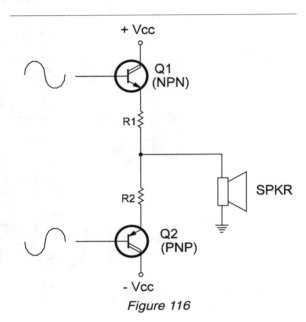

Figure 116

Testing

Darlington transistors can be tested the same way as common bipolar transistors. Using a multimeter, you can measure the resistances between electrodes (base, emitter, and collector) according to the type (NPN and PNP). Refer to the item about bipolar transistors and how to test them. Remember that some multimeters have the transistor test function that can be used with Darlington transistors.

Unijunction Transistor (UJT)

Unijunction transistors are three-terminal devices, but unlike the common bipolar transistors, they have only one junction (as the name suggests). The unijunction transistor or UJT exhibits a negative input resistance that is useful in a number of low-voltage circuits, such as low-frequency oscillators, saw-tooth generators, timers, etc.

As shown by the structure in Figure 117, an N-type silicon bar forms the UJT where two terminals called base 1 and base 2 (B1 and B2) are placed. Between these terminals the device presents an ohmic resistance.

Section II—Electronic Components

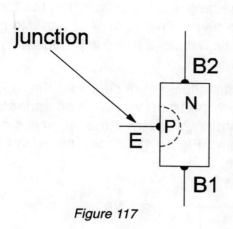

Figure 117

A PN junction is formed in the middle of the N-region where an emitter terminal is placed. In a typical circuit, such as an oscillator, the device is biased as shown in Figure 118.

The N-material placed at the middle of the bar acts as a voltage divider placing a certain voltage in the emitter region. The capacitor placed at the emitter of the transistor charges until the voltage between its plates rises to the value of the emitter voltage plus about 0.6 volts.

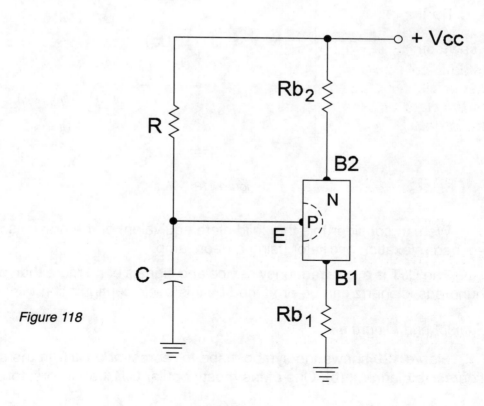

Figure 118

133

When this voltage is reached the device turns on and the resistance between the emitter to Base 1 falls to a very low value. The capacitor can then discharge through this circuit. A high current pulse is produced in this process.

As long as the charge in the capacitor is reduced, and the voltage across it falls to an appropriated value, the transistor turns off and returns to the nonconductive state. The charging process of the capacitor starts again and a new oscillation cycle begins. Figure 119 shows the waveforms of the signals produced by this circuit.

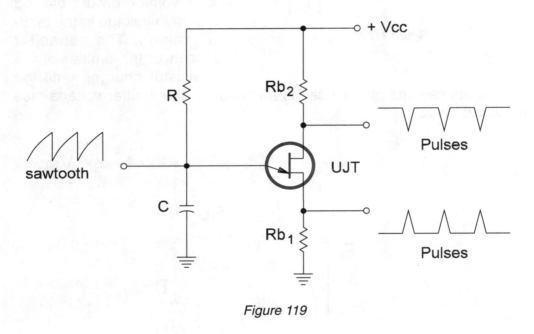

Figure 119

We can consider this the solid-state equivalent of the previously described relaxation oscillator using a neon lamp.

The UJT is a low-frequency device and signals of no more than some hundreds kilohertz can be produced by this basic configuration.

Symbol and Diagram

Figure 120 shows the symbol used to represent a UJT. In the same figure is a diagram, (b). One of the most popular UJT transistors, found in projects published by electronics magazines and even in some kits and appliances, is the 2N2646.

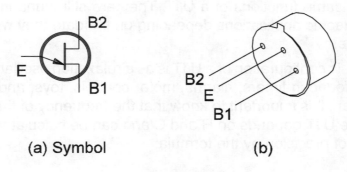

(a) Symbol   (b)

Figure 120

Specifications

As in any other transistor, there is a part number figured on the device's body for its identification. But, when building a project and trying to understand how the circuit works or looking for a replacement type, it is important to know some electrical specifications of a UJT. The most important are listed here.

A. Voltage between bases (Vbb)—the maximum voltage that can be applied between bases. For the 2N2646 the specified value is 35 V.
B. Voltage between emitter and base 1 (Vb1e)—the highest voltage that can be applied between the two indicated electrodes. For the 2N2646 the value is 30 V.
C. Intrinsic standoff ratio (η)—the position of the emitter junction referring to the bases. This value is used to calculate the trigger point as a function of the applied voltages. For the 2N2646 this value is in the range between 0,56 and 0,75.
D. Resistance between bases (Rbb)—the ohmic resistance that can be measured between bases. Typical values for the 2N2646 are in the range between 4.7 and 9.1 k ohms.
E. Peak Pulse Emitter Current (Ie)—the maximum current that can flow between emitter and Base 1 when the transistor is triggered.

Electronics for the Electrician

Where they are found

The UJT isn't a modern device—it was introduced on the market around 1960. Today equivalent circuits using modern devices are used to perform the same functions of a UJT. They are still found in many electronic and electric applications depending on the date they were made or their purpose.

The basic configuration of a UJT is as a relaxation oscillator. This type of circuit is found in timers, alarms, motor controls, toys, and others. For the electrician, it is important to know that the frequency of the signal produced by the UJT depends on R and C and can be calculated with a certain degree of precision by the formula:

$$f = \frac{1}{R \times C \times \ln\left[\frac{1}{1-\eta}\right]}$$

Where
f is the frequency in Hertz (Hz)

R is the resistance in ohms (Ω)

C is the capacitance in Farads (F)

1n is the natural logarithm

η is the intrinsic standoff ratio (about 0.6 for the 2N2646)

How to test

Using a multimeter you can make two tests in a UJT:

1. The resistance between the two bases in any direction is between 4.5 and 12 k ohms for a common UJT.
2. Placing the probes between the emitter and Base1, a high resistance can be measured in one direction and a low resistance in the opposite direction. This means that the junction (acting as a diode) is good.

The best way to test this device, however, is to mount a simple probe circuit like the one shown in Figure 121.

If the transistor is okay, the LED will flash at a rate that can be adjusted by the potentiometer.

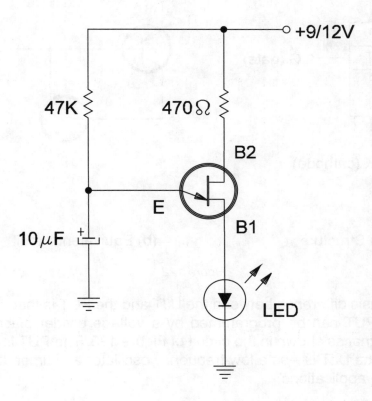

Figure 121

Programmable Unijunction Transistor—PUT

Although this device operates the same way as a UJT, its structure is different. It is a four-layer semiconductor placed in the category of components called thyristors. The structure of a PUT is like the one shown in Figure 122.

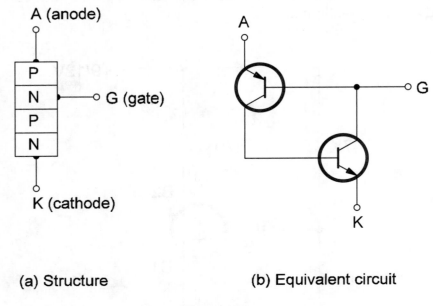

(a) Structure (b) Equivalent circuit

Figure 122

The basic difference between the UJT and the PUT is that the trigger point of a PUT can be programmed by a voltage divider placed in the anode terminal as shown in the circuit of Figure 123. The PUT is used the same way the UJT is—as a low-frequency oscillator and timer. It is found in the same applications.

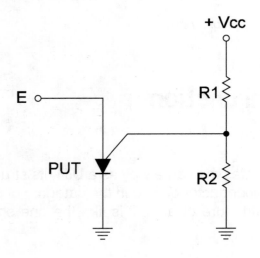

Figure 123

Section II—Electronic Components

Symbol

The symbol of a PUT is shown in Figure 124.

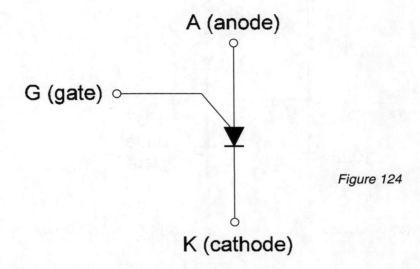

Figure 124

Specifications

The PUT is identified by a part number but there are electrical specifications that can be found in data sheets as the maximum voltage between anode and cathode, and the maximum operation frequency are all similar to the UJT.

Where they are found

The PUT is not a common device in modern applications, but the electrician working with electronics will find it in the same equipment where UJTs are found—timers, motor controls, audio oscillators in alarms, etc. The use of the PUT as a discrete device is not common because manufacturers use the same function in complex integrated circuits that include other functions. Knowing how it works is important because, although the device is not common, the function is.

Testing

The best way to test a PUT is using a circuit where it can drive an LED (Figure 125). If the PUT is okay the LED will flash.

Electronics for the Electrician

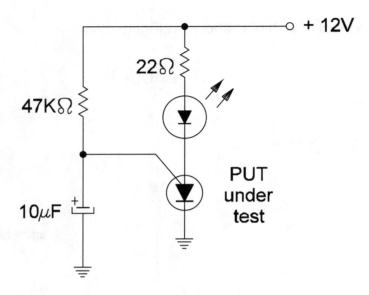

Figure 125

Field-Effect Transistor—FET

The Field-Effect Transistor (FET) is a semiconductor device that can be used in basically the same functions as the bipolar transistor: amplification, switching, or as an oscillator.

The basic FET is formed by a piece of N-type semiconductor material (silicon) where two P regions are formed (Figure 126).

The current through the channel of P material, where the drain (d) and source (s) terminals are placed, can be controlled by a voltage applied to the gate (g) region. The basic difference between a bipolar transistor and the FET is that, unlike the bipolar transistor that controls a collector-emitter current from a base current, the FET controls a current between drain and source from a gate voltage. The bipolar transistor is said to be a typical current amplifier and the FET is a typical voltage amplifier.

Like the basic material where the drain and source are placed (N and P), the FETs can be found in two types: N channel and P channel. This

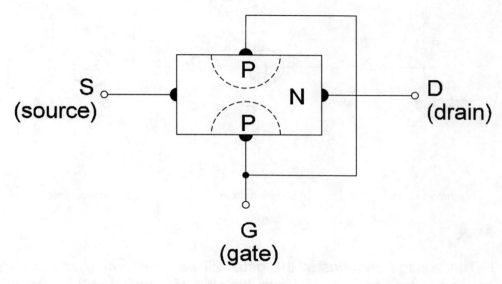

Figure 126

basic FET is referred to as JFET or Junction FET because, between the gate and the channel where the drain and the source are placed, exists a PN junction.

In the basic application the FET is wired as shown in Figure 127.

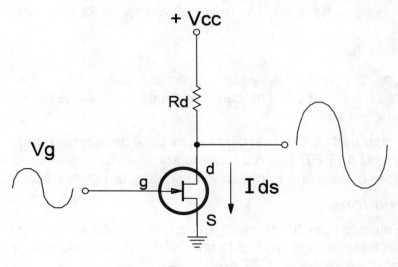

Figure 127

Electronics for the Electrician

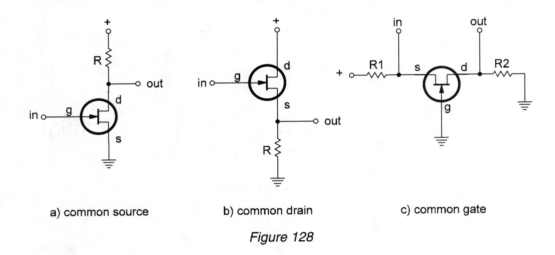

a) common source   b) common drain   c) common gate

Figure 128

The voltage changes in the gate will be transformed into current changes between drain and source. If a signal is applied to the gate, this signal can be taken already amplified from the drain or source of the transistor according to the configurations. The FET can be used in the same configurations as the ones found in bipolar transistor circuits. The common drain, common source, and common gate configurations are shown in Figure 128.

Warning: *FETs are very sensitive devices. The electric charge of your body can destroy the device if you touch its terminals.*

Symbols and Types

Figure 129 shows the symbols used to represent both the P-channel and the N-channel JFET. The common types of these devices are shown in the same figure.

Observe that no external differences exist between a common bipolar transistor and an FET. The only way to know which one is which is by the part number or referring to the schematic diagram of the device.

Specifications

The electric specifications for an FET are important to help the electrician find a replacement type or to know if it can be used in a project. The main specifications for an FET are:

Section 11—Electronic Components

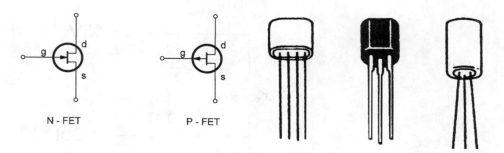

Figure 129

A. Maximum voltage between drain and source (Vds)—the maximum voltage that can be applied to the FET, typically in a range between 20 and 60 volts. This specification can be abbreviated as Vds (max).
B. Maximum drain current—the maximum current that can be controlled by a FET. A common abbreviation in the databooks is Id.
C. Transconductance—the equivalent of the gain of bipolar transistors. It measures the inverse of the resistance between drain and source under certain conditions. The transconductance is measured in Siemens (S). In old publications, the antiquated unit mho (the word "ohm" inverted) is used.
D. Dissipation power—the same as in bipolar transistors and is specified in watts. Small FETs have dissipation powers in a range from 50 to 500 mW.

Where they are found

The JFET or FET is found in circuits where signals must be generated or amplified. Audio circuits and sensor circuits in alarms can use an FET in their input. They can be used in the linear mode as amplifiers or switches like the common bipolar transistor. Figure 130 shows a typical stage where an FET is used as an amplifier for signals from a microphone or any other transducer.

Stages like this can be found in commercial equipment such as alarms, radios, and remote controls.

When replacing an FET it is important to know the type and, if a replacement is used, the placement of the terminals as they can change from type to type.

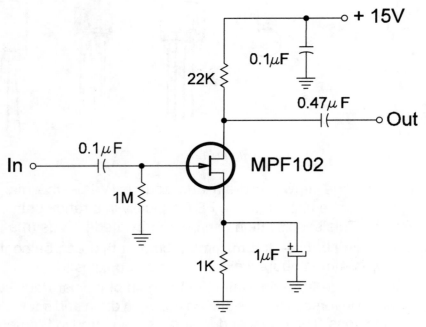

Figure 130

Testing

The ideal for testing an FET is a special meter dedicated to this task. But, even using a simple multimeter, it is possible to detect if an FET has problems. Figure 131 shows how to test an FET measuring the resistance between its terminals.

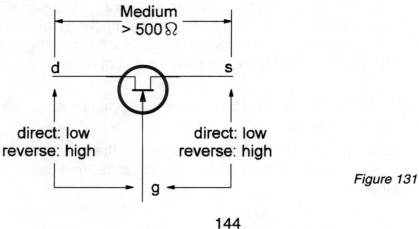

Figure 131

MOSFETS

MOSFET or Metal-Oxide Semiconductor Field-Effect Transistor is a device derived from the common FET but with some changes in the basic structure and also in the electrical characteristics.

As shown by the structure in Figure 132, MOSFETs have a thin layer of metal oxide (silicon oxide) isolating the substrate from the gate region, instead of a junction found in the JFET.

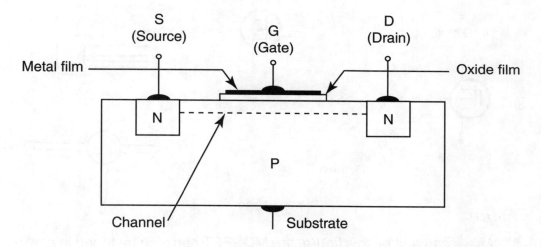

Figure 132

Although there are differences in the structure, the operation of a MOSFET is the same. A voltage applied in the gate can control the amount of current across the device or the current between drain and source. The MOSFET can be used in the same applications where the JFET is found with some advantages. It can be used in signal amplification, signal generation, and many other functions.

Warning: The layer that isolates the gate region from the channel is very thin, making the device extremely sensitive to static discharges. If you have a static charge in your body, touching any terminal of the device can produce a spark between the gate and channel, destroying the isolating layer and the device.

Electronics for the Electrician

Symbols and Types

Figure 133 shows the symbols of the two types of MOSFETs commonly found in electronic circuits. This figure also shows types of these devices.

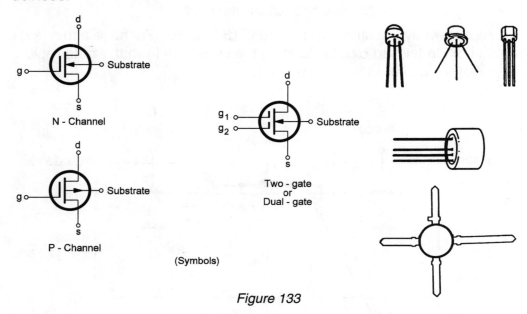

Figure 133

As we can see by this figure, the MOSFET can also be found in a two-gate version (g1 and g2). The amount of current across the device can be controlled independently by the voltages applied to each of the two gates. This device is suitable for applications where two signals must be mixed.

The common MOSFETs are low-power devices and look the same, making it difficult for an electrician to identify them from bipolar transistors. It is necessary to have the schematic diagram, a data sheet with information, or the type (part number on its body).

Specifications

A MOSFET is identified by a part number. By knowing this type number, consulting data sheets, or consulting handbooks it is possible to know its electric characteristics. These characteristics are important in order to find a replacement type. The main characteristics of a MOSFET are as follows:

Section II—Electronic Components

A. Maximum voltage between drain and source—indicated as Vds (max) or only Vds, this is the maximum voltage that can be applied to the device without burning it. Values between 20 and 50 volts are common.
B. Maximum drain current—the maximum current controlled by the device. In common types this current ranges from 10 to 100 mA. It is also abbreviated by Id or Id (max).
C. Transconductance—in the common JFETs, it is the parameter that indicates the amplification capabilities of the device. It is measured in Siemens (S).
D. Dissipation Power (Pd)—as in any electronic device, the MOSFET converts electric energy into heat. The maximum amount of heat generated by the device and transferred to the area surrounding the device is the maximum dissipation power (Pd) given in watts (W).

Where they are found

The MOSFETs are mainly found in high- and low-frequency power circuits as in the input of audio amplifiers and as sensor amplifiers, or in high-frequency low-power circuits like RF, FI, mixer, and oscillator stages in radios, remote controls, receivers, TVs, walk-talkies, etc.

Today the functions of JFETs and MOSFETs can be included in other devices (integrated circuits) as we will see, but in some cases, depending on the equipment, the electrician may find these components.

Testing

The MOSFET can be tested by measuring the resistance between drain and source and also between the gates and the source (Figure 134). However, the best test can be made using a probe circuit that can easily be implemented on a solderless board (protoboard).

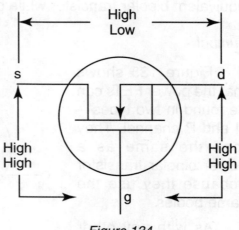

Figure 134

Power FETs

Power FETs are MOSFETs that can be used to control large amounts of current. They are very common today in digital and analog applications replacing the bipolar transistor. Power FETs, or power MOSFETs as they also called, can control currents as large as 50 amperes under voltages of more than 500 volts.

The structure and the operation principle of a power MOSFET are the same as the common MOSFET. The differences are in the size and shape of the silicon chip used when the device is manufactured.

Several technologies are used to produce these devices giving to the POWER-FET transistor names such as VFETs, power FETs, DFETs, and TMOS. The power FETs operate the same way the common MOSFET operates. A voltage applied to the gate can control the amount of current flowing across the device.

High-current devices such as solenoids, motors, electromagnets, relays, heaters, and lamps can be controlled by power FETs in the same kind of stages as the ones that use power bipolar transistors or Darlington transistors. The main advantage found in the use of power FETs is the low resistance presented when conducting a current (called Rds or drain source resistance). Because the amount of heat produced by the device depends on this resistance, power FETs can control much larger currents than the equivalent bipolar transistor while generating less heat.

Symbol

Figure 135 shows that the power FETs can be found in two types— N and P channel. They look the same as a power bipolar transistor because they use the same bodies.

As with any other power transistor, the

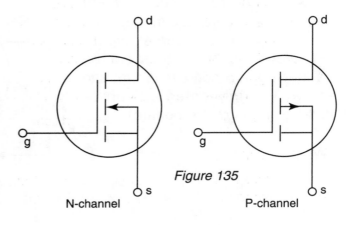

Figure 135

Section 11—Electronic Components

power FETs must be mounted in heat sinks to transfer to the all the heat they generate when working.

Specifications

The power FETs are identified by a part number. A popular series of power MOSFETs and produced by many manufacturers including International Rectifier, Fairchild, and Motorola is the IRF series. All devices of this series have part designations beginning with IRF: IRF640, IR720, IRF630, IRF730, etc.

By knowing the name (part number), it is possible to find the electrical characteristics of a Power FET in handbooks or on data sheets. The principal characteristics are as follows.

A. Maximum voltage between drain and source—the maximum voltage that can be applied to the device without burning it. Common types have voltages ranging from 50 to 1000 volts. This specification can be indicated as Vds (max).

B. Maximum drain current—the highest current that can be controlled by the device and can range from 1 to 50 amperes. This specification is abbreviated by Id or Id (max).

C. Dissipation power—the amount of heat that the device can transfer to the surrounding area. It is measured in watts and can range from 10 to 250 watts.

D. Resistance between drain and source in the on state—the most important of the power FET's specifications. The amount of heat generated by the device depends on this resistance as it operates in more of the applications as an electronic switch. The product of this resistance multiplied by the square of the current gives the generated power. It is easy to see that the lower this resistance, the more efficient the device is when controlling high amounts of current and more heat is generated. This resistance used to be less than 1 ohm in more of the common devices. It is also indicated by rDS or Rds and is given in ohms (Ω).

Where they are found

Power FETs are slow devices used to control high current loads but in low- and medium-frequency circuits. Like the bipolar transistors, the power

Electronics for the Electrician

FETs can be used as a switch (digital mode) or as a signal amplifier (linear mode).

As switches we can find the power FETs controlling solenoids, relays, step motors, and other loads in circuits like the one shown in Figure 136.

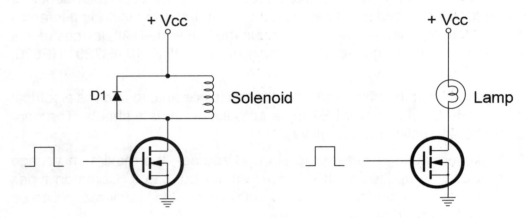

Figure 136

When a positive voltage is applied between gate and the ground, the power FET (P channel) turns on and the current flowing across the device powers on the load.

Another important application of power FETs is in Switch-Mode Power Supplies (SMPS) like the one shown in a block diagram in Figure 137.

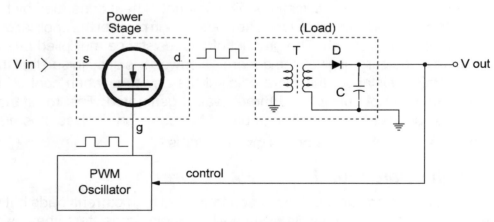

Figure 137

Section II—Electronic Components

In this circuit the power FET acts as a switch that opens and closes very fast in a manner so that the pulses of voltage applied to the load can be controlled in frequency and duration.

If the interval or the width of the pulses changes, the amount of energy they transfer to the load also changes, and with them, the average voltage. By using a feedback circuit it is possible to control the width and interval between pulses to keep constant the average voltage applied to a load. So, in an SMPS exists a circuit that changes the voltage output, controlling the pulse width applied to the power FET to compensate the alterations and keep the output voltage constant. This kind of power supply is found in many types of electronic equipment, such as computers, video monitors, and TVs. In the linear mode, power FETs can be used as audio amplifiers or low-frequency signal amplifiers. A circuit where an audio signal is amplified by a power FET is shown in Figure 138.

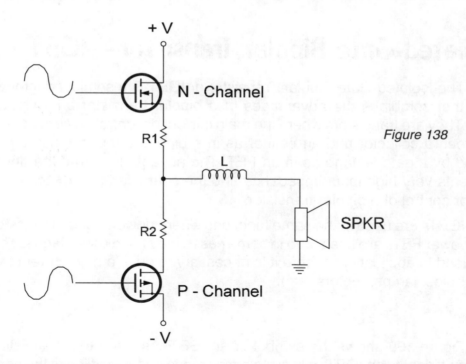

Figure 138

Alarms, which are a common type of audio equipment, can use this type of transistor in their output stages driving the loudspeaker with many watts.

High-fidelity (hi-fi) audio amplifiers that use power FETs are high-quality equipment presenting low distortion levels. The power FETs have characteristics similar to the ones found in the old tubes for amplifying low-frequency signals. Many enthusiasts of the bipolar transistor can't reproduce the sounds with the same quality found in old tube models. But, amplifiers using power FETs can present levels of fidelity closer to those found in the tube amplifiers better than the ones of bipolar transistors.

Testing

A simple multimeter can be used to test a power FET. The resistance between the gate and the other terminals must be very high in a good transistor. If a low resistance is measured the transistor has problems. A low resistance between gate and the other terminals indicates a transistor with losses or a short.

Isolated-Gate Bipolar Transistor—IGBT

The Isolated-Gate Bipolar Transistor or IGBT is a semiconductor device that combines the advantages of a bipolar transistor and a power FET. They are transistors where the main current or controlled current flows between a collector and an emitter as in a bipolar transistor, but is controlled by a gate voltage as in an FET. The advantage is that the device presents very high input impedance and the characteristics of commutation or control of a bipolar transistor.

IGBTs are used in the same functions where power bipolar transistors and power FETs are used, and in some cases they have advantages. They are used in applications where it is necessary to drive a high current load (solenoids, lamps, motors, etc).

Symbol

Figure 139 shows the symbol of an IGBT. As with other transistors, we can't identify an IGBT only by its appearance because it looks the same as bipolar transistors and power FETs. It is necessary have a part number or a schematic diagram of the equipment.

Section 11—Electronic Components

Specifications

IGBTs are identified by a part number. From the part number, the electrical characteristics can be discerned. The main characteristics are as follows.

A. Maximum voltage between collector and emitter—the maximum voltage that can be applied to the device. It is also represented by Vce (max).

B. Maximum collector current—the maximum current that can be conducted by the device. It is also abbreviated by Ic (max).

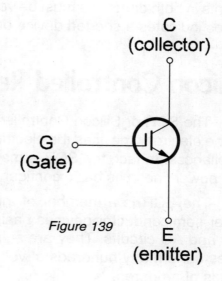

Figure 139

C. Dissipation power—the maximum amount of power that the device can convert in heat and transfer to the surrounding area. It is given in watts and abbreviated as Pd (max) or Pd.

Where they are found

IGBTs are components widely found in industrial applications. The advantages of the use of this device in the control of high-current loads in industrial machines, such as motors, solenoids, electromagnets, and heaters, make them very commonly used in these types of equipment.

In many cases, the IGBT can directly replace a power FET or a Darlington bipolar transistor. The most the technician must do in many cases is change the value of some resistors in the circuit.

As IGBTs are new devices they are not very common in electric or electronic applications more than 10 years old. In home appliances, such as dishwashers, clothes washers, and others that use inductive loads, the IGBT can be found as the control element.

Testing

The simplest way to test an IGBT is measuring the resistance between the terminals. The resistance between the gate and the other elements in both directions must be very high. A low resistance in these measures indicates a shorted device or one with losses.

Silicon Controlled Rectifier—SCR

The SCR or Silicon Controlled Rectifier is one of the most important of the electronic devices the electrician will find in electric installations and appliances. In fact, the SCR is made to control devices plugged in to the AC power line in its basic configuration.

The SCR is a member of the thyristor family. Thyristors are four-layer semiconductor devices basically intended for applications of control and AC circuits. They are solid-state switches, can operate at voltages up to many hundreds of volts, and can handle currents up to hundreds of amperes.

As the SCR is very important to the electrician who wants to know how electronics devices are used in electric installations, more space than the usual dedicated to other components is given to their description and applications. So, what is an SCR, starting from its structure?

Figure 140 shows the structure of an SCR and the equivalent circuit formed by two complementary transistors. It is a silicon semiconductor with a PNPN structure.

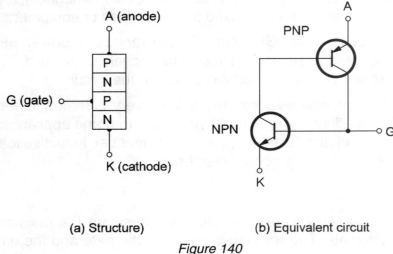

(a) Structure)   (b) Equivalent circuit

Figure 140

Section II—Electronic Components

In the basic operation the anode (A) is made positive and the cathode (K or C) negative. The load is placed between anode and the power supply. If no voltage is applied to the gate (G) the two transistors remain off and no current can flow across the load. But, if enough voltage is applied to the gate, it can make the NPN transistor conductive and a current begins to flow between its collector and emitter.

This current is the one that comes from the base of the PNP transistor biasing on this component. The current that now begins to flow by the NPN transistor acts as a feedback to the NPN transistor as it also flows through its base. The feedback makes all the currents in the circuit increase speed and in a few microseconds the circuit reaches a saturated state. The current between the collector of the PNP transistor and the emitter of the NPN transistor is now at its maximum value. After completion of this very fast process, even if the initial voltage that started all this is no longer present, the device remains conductive.

In a real SCR the process is the same: a positive voltage applied in the gate triggers on the device that conducts the current from the anode to the cathode. Even if the voltage disappears once it is triggered, the device is kept in the on state. To trigger it off there are two possibilities:

1. Turn off the power supply for a moment. This makes the device current fall to zero.
2. Reducing the current flowing through the device to a value that cuts the feedback process. A short between anode and cathode, reducing the voltage to zero (Figure 141), can be created to trigger the SCR off.

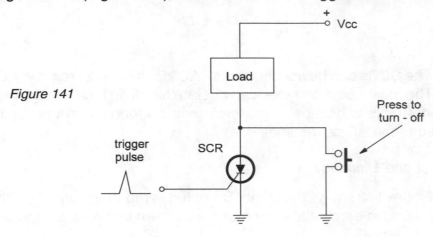

Figure 141

Electronics for the Electrician

Notice that the SCR acts as a diode that can be made conductive or not conductive because the current flow only can flow in one direction. This explains its symbol—a diode with a gate.

Real SCRs can be triggered by voltages of about 1 volt and currents ranging from 100 microamperes for the most sensitive to some miliamperes for high current types and less sensitive types.

If placed in an AC circuit like the one shown in Figure 142, when turned on the SCR can conduct only the positive semicycles of the AC power line voltage.

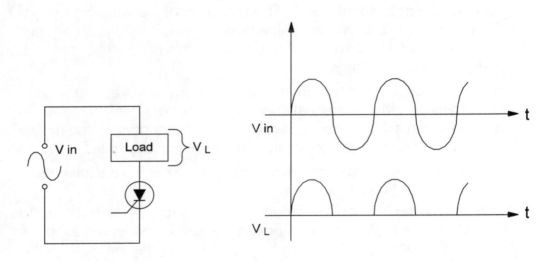

Figure 142

The SCRs can be used to control AC loads directly from the AC power line. The most common types can operate handling high voltages and high currents. This is because this device can be found in many applications related to the AC power line.

Symbol and Diagram

Figure 143 shows the symbol adopted to represent an SCR. The main types found in electronic equipment are shown in the same figure.

Section 11—Electronic Components

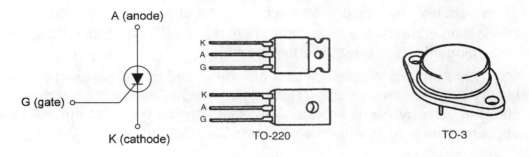

Figure 143

Although small SCRs designed to control low currents still exist, more of them are power devices including resources to place them on large heat sinks. The tab or the metal body of the SCRs used to be connected to the anode terminal.

Specifications

The SCRs are identified by a part number given by the manufacturer on the body. From this part number, looking in a handbook, in a databook, or on a data sheet it is possible to know the electrical characteristics.

When using any SCR it is important to know the following characteristics:

A. Maximum voltage between anode and cathode—the maximum voltage that can be applied to the SCR when it is in the off state. Typical values range from 50 to more than 1000 volts for common types. This specification can also be indicated as Repetitive Peak Off voltage and abbreviated by Vdrm or as Repetitive Peak Reverse Voltage and abbreviated by Vrrm.

B. Maximum current—the maximum current that the SCR can hold when in the conductive state. Common SCRs found in AC-powered appliances are specified to control currents in the range from a fraction of an ampere to more than 50 amperes.

C. Dissipation power—When an SCR is on, independently of the current it is conducting, the voltage between anode and cathode is about 2 volts. The product of this voltage by the current gives the dissipation of the

SCR. The maximum dissipation of an SCR is important to know in order to calculate the size and shape of the heat sink.

D. Holding current—the lowest current that the SCR can conduct without turning off. This current is abbreviated by I_H.

Other important specifications are the speed of the device given by the value referred to as Critical Rate of Rise of Off State Voltage and measured in volts by microseconds. This value shows how fast the voltage applied to the load increases when the SCR is turned on.

Where they are found

SCRs are found in several applications involving AC power control. Some devices where SCRs can be found are motor controls, timers, dimmers, alarms, solenoid drivers, and many home appliances.

The most common application of an SCR as a power control is the one shown in Figure 144.

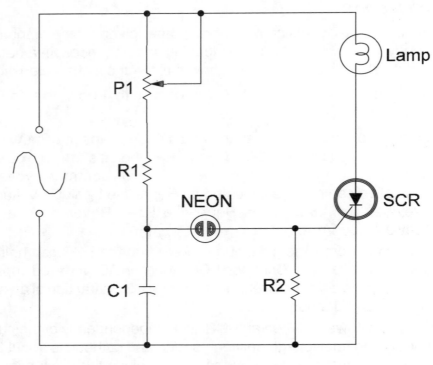

Figure 144

Section II—Electronic Components

This circuit is found in commercial dimmers like those installed to replace common on/off switches. It is also used to control the brightness of incandescent lamps, in fans as a speed control, and in heaters as a temperature control.

It is very important for an electrician to know how this circuit works because its operation principle is applied to a large number of common electric appliances that have a power (speed, temperature, etc) control.

Look at the waveform of the AC power line voltage shown in Figure 145.

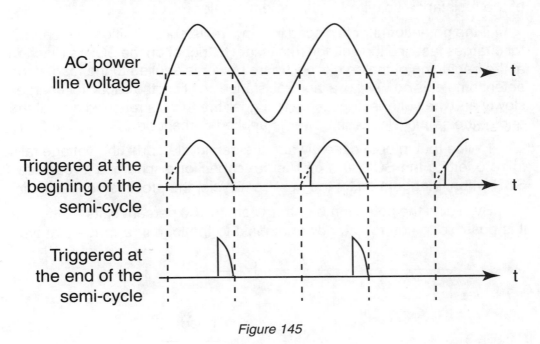

Figure 145

The voltage changes make the current flow forward and backward across a load determining the amount of power absorbed by it. This can be reduced if we cut the semicycles of the AC power line voltage in different points.

For example, if the semicycles are cut at the end, only a small part of the AC power line voltage passes and is applied to the load. If it is a lamp it will present a low brightness. If it is a motor it runs at low speed. If it is a heater the temperature is lower. If the semicycle is now cut at the beginning, a larger part of the power passes and is applied to the load. If it is a

Electronics for the Electrician

lamp, the brightness is increased, a motor runs at a higher speed, and a heater is warmer.

Using an appropriate control, we can slide the point where the semicycles are cut from near zero to 100 percent and thus control, with a great precision, the amount of power applied to a load.

The circuit shown in Figure 144 makes this control. At the beginning of a semicycle of the AC power line the voltage applied to the capacitor C increases in speed depending on the value of the component and the value of the resistance adjusted in the potentiometer P1.

If the potentiometer is placed in a low-resistance position, the capacitor charges fast and the voltage necessary to trigger on the SCR is reached at the beginning of the semicycle. More power is applied to the load. If the potentiometer is now slid to a high-resistance position, the capacitor charges slowly and the voltage necessary to trigger the SCR is reached only at the end of the semicycle. Less power is applied to the load.

Notice that in both cases, when the semicycle ends, the voltage falls to zero letting the SCR turn off. As the capacitor discharges through the SCR when it triggers on, in the next semicycle, the process begins again.

By sliding the potentiometer from zero to 100 percent of its resistance it is possible to control the power applied to the load in a range that typi-

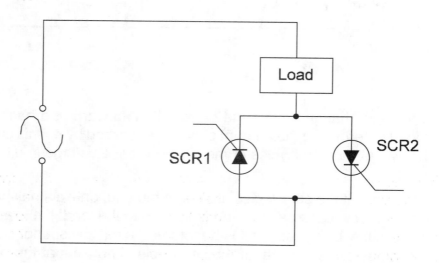

Figure 146

Section II—Electronic Components

cally goes from near 0 to 95 percent of the maximum power. A little problem must be considered in this circuit: the SCR is a unilateral device and the AC loads must receive both semicycles to reach their normal maximum power. How can this problem be solved?

In some cases the circuit can operate only with the positive semicycles of the AC power supply line, and the SCR can be used. In other cases, two SCRs wired in an anti-parallel mode are used, as shown in Figure 146. One SCR conducts the positive semicycles and the other conducts the negative semicycles of the AC power line voltage.

Another solution to the problem is in the use of a diode bridge (rectifier) that converts the AC power line voltage into a DC pulsed voltage, or a voltage formed only by positive voltage semicycles as shown in Figure 147.

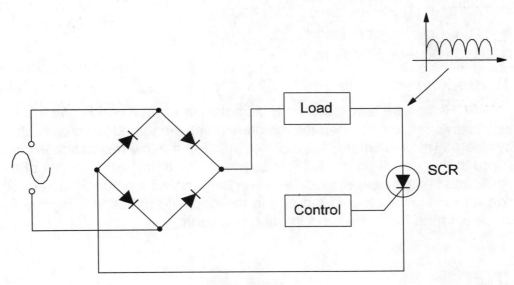

Figure 147

But, the best solution is in the use of another device of the thyristor family—the triac. The triac can be considered as two SCRs inside a unique case that can be used to control both semicycles of the AC power line voltage. The SCR is not used only as a power control. Many other configurations are possible including timers, oscillators, alarms, automatic switches, and protection circuits (crowbar).

The 106 Series

One of the most popular series of SCRs found in many electronic circuits is the 106 series. It is formed by low-cost general-purpose SCRs that can control currents up to 3 or 4 amperes (3.2 volts average for the TIC106) in circuits having voltages in the range between 50 and 1000 volts. The devices of this series are very sensitive; they can be triggered from currents as low as 60 uA (typical). The part number of the SCRs of this series begins with a group of letters identifying the manufacturer followed by 106, and a figure indicating the maximum operating voltage.

Some examples follow.

- MCR106-B (Motorola - the B indicates it is for 200 V)
- TIC106-6 (Texas Instruments for 400 V)
- IR106 (International Rectifier)
- C106 (General Electric)

Testing

SCRs can be tested with a multimeter. In normal conditions the device is an open circuit when the resistance between anode and cathode is measured (the multimeter will indicate infinite). If a low resistance is measured in this condition the SCR is shorted. The other test consists of the verification of the gate-cathode junction that acts as a diode. In one direction a low resistance is found and in the opposite direction a high resistance is found if the device is in good condition.

Triac

The Triac is another important device of Thyristor family. As demonstrated, SCRs are unidirectional thyristors meaning that they can control the current only in one direction—from anode to cathode. Unlike the SCR, triacs are bidirectional and can conduct the current in both directions. Compare a triac to two SCRs connected in inverse parallel (anti-parallel) and placed inside a single three-terminal case as shown in Figure 148.

Section II—Electronic Components

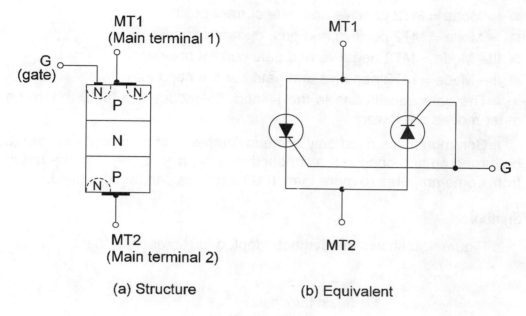

(a) Structure   (b) Equivalent

Figure 148

The triac is indicated to applications where AC voltages must be controlled meaning that a large range of applications in domestic and industrial installations includes this component. So, it is another device with a large degree of importance for the electrician. In basic applications the triac is connected in a circuit as shown in Figure 149.

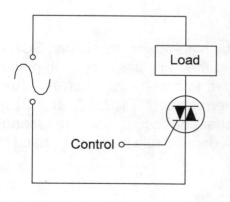

Figure 149

Like the SCR, the triac can be used as a power switch controlling a large amount of alternating current from the main terminals (MT1 and MT2) from a low voltage applied to the gate. Unlike the SCRs that need only positive voltages applied to the gate to be turned on, triacs can be triggered either by a positive or a negative gate signal, regardless of the polarities of the main terminal voltages. This means that the device can be triggered by four modes, indicated as follows:

Electronics for the Electrician

- I+ Mode = MT2 positive and gate current positive
- I− Mode = MT2 positive and gate current negative
- III+ Mode = MT2 negative and gate current positive
- III− Mode = MT2 negative and gate current negative

The gate sensitivities in the I+ and III− modes are high and in the other modes are lower.

Common triacs need only some miliamperes of current in the gate to be turned on and once in the conductive state, high currents in the range from some amperes to more than 1000 amperes can be controlled.

Symbol

Figure 150 shows the symbol adopted to represent a triac.

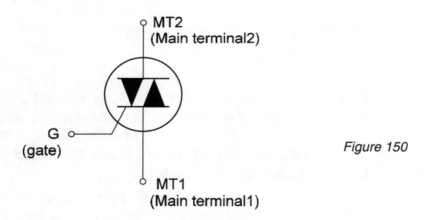

Figure 150

The body is the same as that used by high-power transistors, SCRs, and many other components. Notice that they are basically high-power solid-state switches and must be mounted in heat sinks when handling high currents. The identification of the terminals (MT1, MT2, and G) is made using the information of the data sheet or documents of the manufacturer as it changes from type to type. A general rule is that MT1 used to be connected to the case or to the tab.

Specifications

The triacs are identified by a part number displayed as a figure on the body, the same way the SCRs, transistors, and many other solid-

Section II—Electronic Components

state devices are. From the part number, using a data sheet or other document from the manufacturer, it is possible to find the electrical characteristics of the device. The inverse procedure is valid as well; from the characteristics, it is possible to find a triac suitable to an application by the part number.

The main specifications to be observed in a triac are:

A. Maximum voltage—the maximum voltage that can be applied to the triac in the off state. They are indicated also as the Repetitive Pulse Reverse Voltage (Vrrm) or Repetitive Peak Reverse Voltage (Vrrm) and can be typically in the range between 50 and more than 1000 volts for common types.

B. Maximum current—the maximum current that can be controlled by the device is indicated as the Continuous On State Current or average On State and can be in the range of 1 ampere to more than 100 amperes for common types.

C. Trigger current and voltage—the current that triggers the triac on is indicated as the Gate Trigger Current and abbreviated by Igt. Typically this current is in the range between some microamperes and 100 mA depending on the size of the device. The trigger voltage is also indicated as Vgt and is typically in the range between 0.8 and 1.5 volts for common types.

D. Dissipation power—When the triac conducts the voltage across, it is typically between 1.7 and 2.2 volts. This voltage times the controlled current or the load current gives the dissipation power of the device. The maximum dissipation depends on the device.

Where they are found

Triacs are found in all the modern applications where a speed, temperature, or power control must be made from an AC power line voltage. We can say that triacs are solid- state switches with a large range of applications in AC control.

Dimmers, speed motor control of electric appliances, heaters, and many other circuits use triacs as a central element.

Figure 151 shows a typical power control or dimmer for loads powered from the AC power line using a triac.

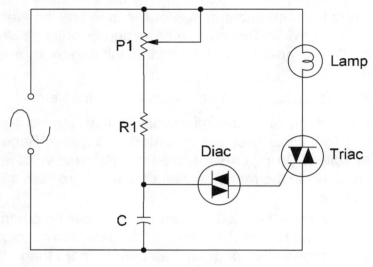

Figure 151

The operation principle is the same as the power control using SCRs previously described. The only difference is that the capacitor charges with both semicycles of the AC voltage and the triac triggers with them conducting both to the load.

The waveforms in the circuit are shown in Figure 152.

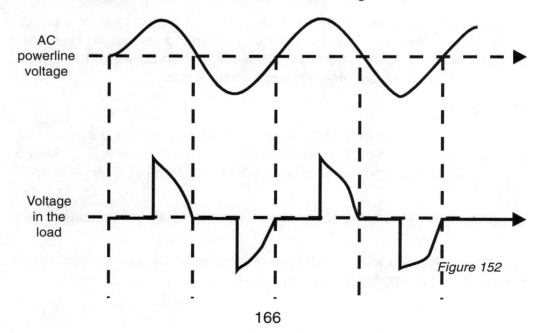

Figure 152

Section II—Electronic Components

The diac is an auxiliary element used to help the triac trigger with the correct voltage adjusted by the potentiometer. Circuits like this are used in common incandescent lamp dimmers replacing common on-off switches. The speed control of motors used in electric appliances, such as drills and fans, use a circuit like this.

Testing

The best way to test a triac is using an appropriated circuit for this task. The multimeter can only reveal if a triac is shorted. If a low resistance is measured between MT1 and MT2 in a triac out of a circuit it is because it's shorted. When replacing a triac by an equivalent type, the specifications (voltage and current) must be equal or greater than the original.

Electromagnetic Interference (EMI)

Triacs and SCRs are high-speed power switches. Their turn-on and off times are very low, only a few microseconds, and this can cause problems with Electromagnetic Interference.

EMI or RFI (Radio Frequency Interference) is the interference caused by undesirable radio signals produced by devices like thyristors affecting the operation of many sensitive pieces of electronic equipment.

When an SCR or triac is used in a circuit as power control, the high-speed changes of the current in a load result in the generation of a series of harmonically related radio-frequency signals. The magnitude of the fundamental signal is proportional to the magnitude of the controlled current, and in a larger part of the cases, can be great enough to cause interference in AM radios and many other circuits that operate with low- and medium-frequency radio signals.

Specifically, the interference is higher when the SCR or triac controls inductive loads in motors, solenoids, magnets, etc. The snap action of the triac or SCR makes the current oscillate in the circuit generating a strong electromagnetic field.

A protection circuit that minimizes the interference is shown in Figure 153. The capacitor and the resistor form a damping circuit reducing the current oscillation and the electromagnetic interference.

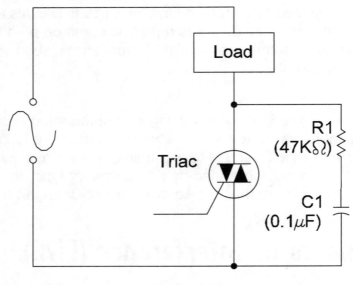

Figure 153

The noise in an AM radio receiver produced when a circuit using triacs or SCRs is turned on or the small dots and points in a TV screen when the same device is turned on are examples of the interference process.

The interference can reach the radio receivers, TVs, or other sensitive devices two ways. The first way is for it to be irradiated directly into the air as a radio signal (Figure 154).

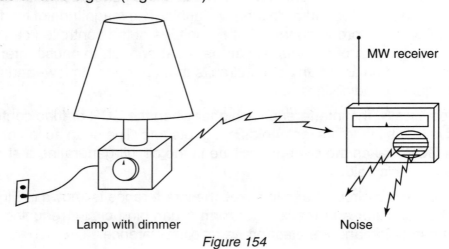

Figure 154

Section II—Electronic Components

The undesirable signal, in general, is very weak and only causes problems if a radio or other type of sensitive equipment is placed near the source of interference. It is also concentrated in the low band of the radio spectrum as shown in Figure 155, interfering much more with power in the LF (low frequency) and MW (medium wave) band.

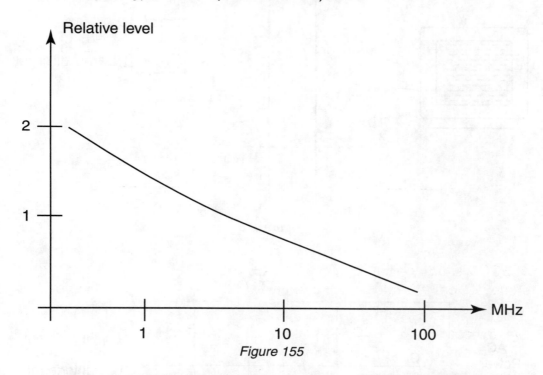

Figure 155

Radio receivers operating at higher frequencies, such as FM receivers, VHF receivers used in remote controls for garage doors, and cellular telephones aren't much affected by this interference not because of lower power concentrated in this band, but also because of their operation principles.

Secondly, the signal reaches the interfered equipment by the AC power line, as shown in Figure 156.

To avoid the interference in this case, use an L-C filter like the one shown in Figure 157.

The electrician can have this problem many times. When installing a dimmer or other device using SCRs and triacs in any domestic electric installation, a TV or an AM radio placed near it or powered from the same

Electronics for the Electrician

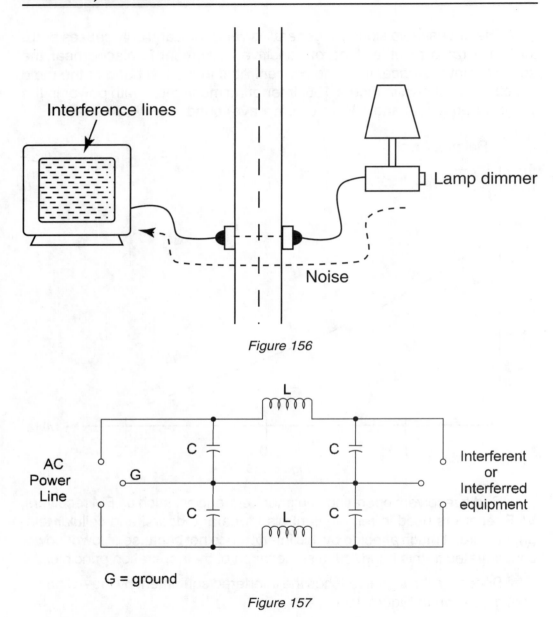

Figure 156

Figure 157

G = ground

AC power line can interfere. A filter like the one shown in Figure 157 placed between the dimmer or the interfering equipment and the power source can solve the problem or reduce the interference to a comfortable level.

Section II—Electronic Components

Diac

The DIAC is another semiconductor device of the thyristor family. It is also a four-layer component with special trigger characteristics. The diac is the semiconductor equivalent of the neon lamp. When the voltage across a diac reaches a certain value, the device triggers on, passing from a high-impedance state to a very low-impedance state. The high-speed triggering characteristics of this component make it ideal as a trigger element for circuits using triacs.

Symbol

Figure 158 shows the symbol used to represent a diac. Notice that the diac is a nonpolarized component because the triggering characteristics are the same in both directions.

Figure 158

Specifications

Associated to a part number, the diacs have a trigger voltage. Typical values are around 35 volts.

Where they are found

Although many applications exist with diacs as a central element of a circuit, the most common is as a trigger element of a triac as in the dimmer. Speed controls for electric motors and dimmers are examples of circuits of where triacs are found, the triggering devices are the diacs.

Testing

The multimeter can't be used to test diacs as they don't have enough voltage to trigger the multimeter and when turned off, they act as an open circuit. The best way to test a diac is using a circuit. The multimeter can only reveal if the diac is shorted when presenting a very low resistance.

Quadrac

The quadrac is a device formed by a diac and a triac in the same body. It is also used as a power control for incandescent lamps, heaters, motors, and other inductive loads the same way the triacs and diacs are used.

Symbol

Figure 159 shows the symbol of a quadrac. They use the same body as power transistors.

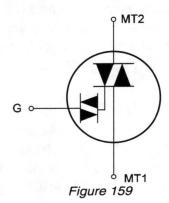

Figure 159

Specifications

The specifications of a quadrac are the same as a triac and a diac. The maximum voltage, maximum current, and trigger point are indicated by the manufacturer. A part number or type number is used to design the component.

Where they are found

Quadracs are not common today. They can be found in old motor controls, lamp dimmers, and some electric appliances where the speed of a motor, temperature, or any amount of power applied to an AC load must be controlled.

The basic application is the same as a triac in power controls as explained before.

Testing

The multimeter can only reveal if the quadrac is shorted. Low resistance between any of the three terminals indicates a problem.

Silicon Unilateral Switch—SUS

The Silicon Unilateral Switch is a device of the thyristor family and is used as a trigger element in circuits using SCRs. The SUS is an SCR with an anode gate and a built-in zener diode, as indicated by the equivalent circuit shown in Figure 160.

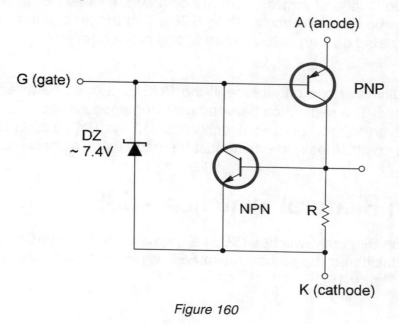

Figure 160

The zener diode determines the triggering voltage, which is normally around 8 V. But, by connecting a lower-voltage zener between the gate and the cathode it is possible to reduce the triggering voltage.

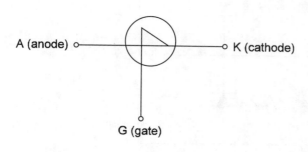

Figure 161

Symbol

Figure 161 shows the symbol of an SUS. It looks the same as a common low-power transistor because they use the same bodies.

Specifications

The main specification of an SUS is the triggering voltage or the internal zener voltage. Other specifications are the maximum current and the maximum voltage supported by the device. Manufacturers identify SUS by part numbers.

Electronics for the Electrician

Where they are found

Silicon unilateral switches are not common devices today. They can sometimes be found in circuits using SCRs as a trigger element. The applications using only an SUS as core of any project are rare.

Testing

A multimeter can only reveal if the SUS is shorted between anode and cathode. The resistance between gate and anode is low in one direction and high in the opposite direction due to the PN junction between these two points. Ideally, the test must be made using a special circuit.

Silicon Bilateral Switches—SBS

Silicon Bilateral Switches (SBS) are devices of the thyristor family. It can be considered the same as two SBS wired in anti-parallel mode as shown in Figure 162.

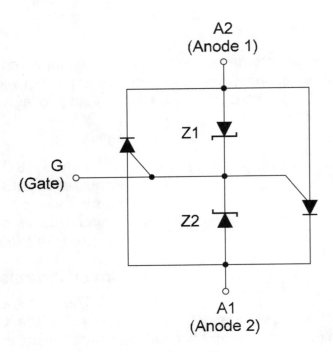

Figure 162

They are used as triggering elements for circuits using triacs. The operation is the same as the SUS; the internal zener diodes determine the trigger point of each SUS and the gate can be used to alter this value. External zener diodes can be wired between the gate and the two other terminals (A1 and A2) to lower the trigger point of the device.

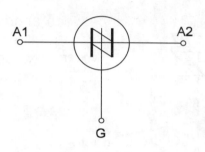

Figure 163

Symbol

Figure 163 shows the symbol adopted to represent the SBS.

Specifications

A part number is adopted by the manufacturer to indicate the device is an SBS. However, only with a data sheet or other manufacturer's document is it possible to access the electrical characteristics of the device, which are the maximum current and voltage between A1 and A2 and the voltage of the internal zener diodes.

Where they are found

The SBS isn't a common device in modern equipment. They can be found in some equipment using SCRs as motor controls, timers, alarms, sensor interfaces, etc.

Testing

The multimeter can only reveal if an SBS is shorted. The operation test must be done using an appropriated circuit.

Liquid Crystal Displays—LCD

Liquid Crystal Displays or LCDs are found in a lot of modern electronics. The liquid crystal displays can display information like the results of arithmetic operations in calculators; time in clocks, timers and watches; programmed operations in dishwashers, clothes washers, and microwave ovens, as well as many other modern applications.

The liquid crystal displays use the electrical properties of the molecules of certain substances that can change their positions when an electric field acts on them as shown in Figure 164.

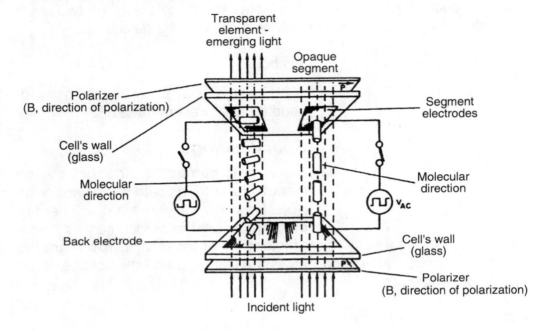

Figure 164

With no electric field present in the substance the molecules are oriented in a manner that lets the light pass through the liquid. The substance is now transparent. When the electric field is applied, the position of the molecules changes making them cut the light. The substance, under this condition, becomes opaque.

If transparent electrodes with the pattern to be displayed are placed inside a can filled with a liquid presenting this property, when a voltage is applied in any part or electrode, this part becomes opaque projecting the corresponding image as shown in Figure 165.

The simplest LCD display is the seven-segment type used to display numbers from zero to nine according to excited segments. This type of display is found in clocks, watches, calculators, and timers, just to name a few.

Section II—Electronic Components

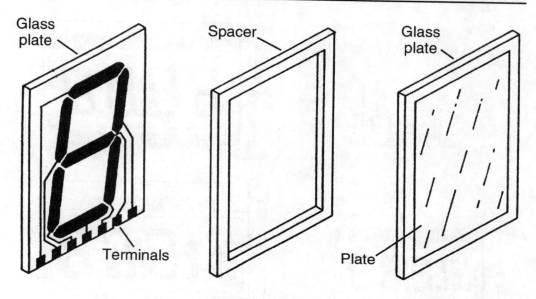

Figure 165

More complex displays include other symbols that can be energized with special signals such as figures in mini games, audio equipment, and radios.

The LCDs need special circuits to become active. The circuit not only recognizes what places of the display must be triggered to show the desired figure, but then also applies an AC voltage. If a DC voltage is used, an electrolysis phenomenon can occur causing the deterioration of the LCD.

The LCD drivers have special characteristics and are included in the group of the ICs (integrated circuits) described later.

Symbols and Views

Figure 166 shows the symbols used to represent an LCD. The same figure shows views of some of these displays.

There isn't any special symbol because many variations can be found in this component based on the application. Normally a representation that corresponds as near as possible from the final end is adopted and easily recognized by anyone who is reading a schematic diagram or working with the circuit.

Electronics for the Electrician

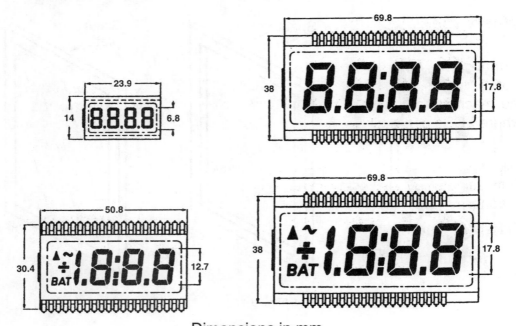

Dimensions in mm

Figure 166

Specification

The LCDs are indicated by a part number in most cases. We can separate LCDs in two groups: general-purpose types and special or custom types.

General purpose are those that present generic symbols (normally seven segments or matrix types) that can be used to present information for the projection. The same type of display can present several types of informational messages.

Specific types or custom types are those that are prepared for a determined manufacturer that creates a special configuration of information for a product. For example, the displays that have the symbol of a company or a figure representing the parts of equipment that are activated at any given time, are examples of these displays.

The electrical characteristics are not so important in this case as they always are driven by a special circuit (usually an integrated circuit).

Where they are found

LCDs are found in any application where numbers, alphanumeric information, or another type of information must be displayed. Watches, clocks, timers, microwave ovens, radios, telephones, car panels, industrial machine control panels, and facsimile machines are examples of places where LCDs can be found.

Some important information to know about the use of these devices is that they have a power consumption many times lower than the equivalents using LEDs so they are especially suitable for battery-powered applications.

Testing

No simple practical electrical test can provide any information about an LCD. The best way to test is using a specific circuit.

Tubes

Although integrated circuits (IC) and transistors are replacing the tubes in most modern applications, there is still old equipment in use today based on these components. To understand how the tubes work and are used, it is interesting to start from the simplest type: the diode.

A vacuum exists inside a glass tube with some electrodes. In the case of a diode, a tungsten filament is placed inside (in some cases near an electrode called a cathode). Also nearby (without touching) is another electrode called an anode. To put the tube to work a positive voltage is applied to the anode, the cathode is connected to a negative voltage, and the filament is heated by a low-voltage source as shown in Figure 167.

When the filament and cathode are heated, electrons are emitted causing current flows between the filament (cathode) and the anode. If the anode is negative and the filament is positive no current can flow across the tube. In a tube like this, the current can flow only in one direction, such as in a semiconductor diode.

If we now place a metallic grid (Figure 168) between the anode and the cathode we can observe new electrical properties in the device.

Electronics for the Electrician

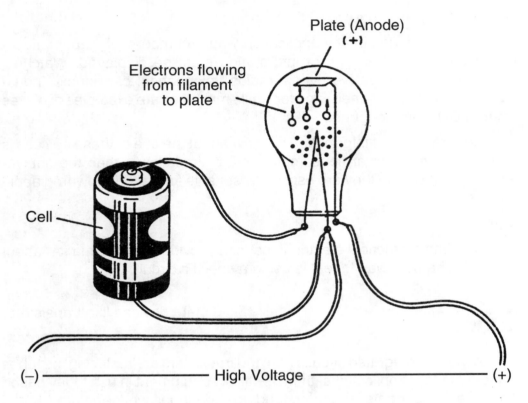

Figure 167

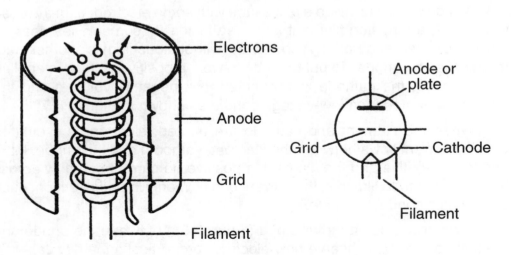

Figure 168

Section 11—Electronic Components

A voltage applied to this grid can control the electron flow between cathode and anode. For example, if the grid is made negative the electrons are repelled and no current can flow. On the other hand, if the grid is made positive, the electrons are attracted and accelerate to the anode.

The amount of current between anode and cathode can be controlled with precision by the voltage applied to the grid. This device, called a triode, can be considered as an equivalent of the transistor and is used for the same purposes—as a switch or an amplifier.

New tubes were developed from the triode configuration by adding new internal elements. The tetrode (four elements), pentode (five elements), hexode (six elements), and heptode (seven elements) are all examples of tubes that can be found in old equipment like radio receivers, audio amplifiers and TVs.

Although the tubes are very efficient in some applications and even used in some of them until recently, they have some disadvantages when compared with the equivalent modern solid-state devices like transistors.

The tubes need to be heated to operate and this implies the consumption of large amounts of energy. Another problem is that all the heat can be transferred to the surrounding area. The tubes also need high voltage for their operation. Although common solid-state transistors operate with low voltages ranging from 1 to 30 volts in typical applications, in the same applications, the tubes need voltages in the typical range between 150 and 800 volts! Portable equipment using tubes is almost totally impractical.

Symbols and Examples

Figure 169 shows the symbols for many types of tubes found in old equipment (and even some modern) and two examples.

In the old types found in some military equipment, metal enclosures replaced the glass and the multiple tubes containing more than one configuration inside a body (a triode and a pentode, for example).

Specifications

The tubes are identified by a part number. There are two main codes used in their specifications. One of the codes uses a number followed by a

Electronics for the Electrician

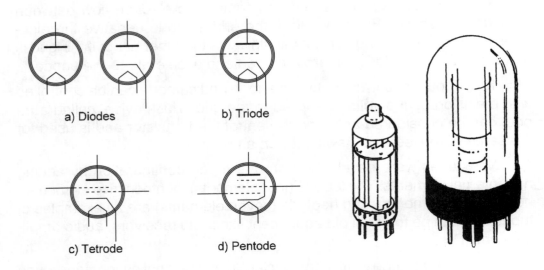

Figure 169

group of letters and another number. The first number is the filament voltage, or the voltage that must be applied to the filament to heat the tube. The second and third figures indicate the specific function. For example, a 12AX7 is a double triode tube that has a 12 V filament.

Another way of identification is the European code, where a group of letters indicates the function of the tube and the final number the specific type. For example, the tube ECC83 is a double triode (CC) (equivalent to the 12AX7!).

The best way to identify each tube is by consulting a manual. (Prompt® Publications has an excellent replacement guide for tubes, *Tube Substitution Handbook*, Sams 61036, ISBN 0-7906-1036-1.)

Where they are found

The electrician will not often find tubes in modern electric installations. The tubes can be found in older electronic equipment, especially radios and TVs. The tubes can be used in the same configurations as the transistors, as shown in Figure 170.

Notice the tubes need, in addition to the low voltage to heat the filament, a high-voltage source for the anode. This high voltage ranges from 150 to 800 volts in common equipment.

Section II—Electronic Components

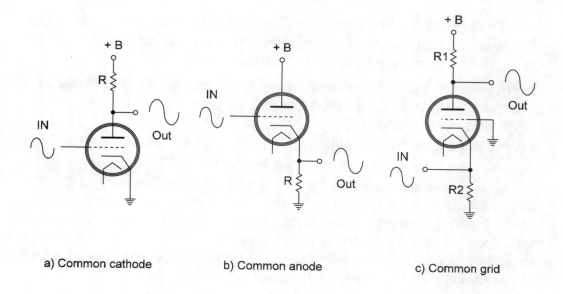

a) Common cathode b) Common anode c) Common grid

Figure 170

Although in some cases it is not a good idea to try to restore old tube equipment, if you like it (perhaps for sentimental value), many dealers have all types of old tubes for replacements.

Testing

When the filament is open (burned out), the tube doesn't function any more. In most of the cases, the malfunction of any piece of tube equipment is caused by this problem. The advantage of the tubes for the technicians is that it is easy to see if a tube is burned out because when the equipment is turned on, there is no orange or red light emitted by its filament. A visual inspection can reveal problems in this case. But, this is not conclusive; some equipment exists where the filaments are wired in series. So, if one of them burns out, all the others are not powered and all the tubes are off.

A multimeter can be used to test the filament of a tube. The operational test needs special circuits or test equipment.

Integrated Circuits

The Integrated Circuit can't be considered a component, but a collection of components housed in a single chip and installed inside a unique case. The basic idea originated from these questions:

1. Why do we have to manufacture all the components of a circuit in a separated process and then assemble them together to create the desired configuration?
2. Why not make them together in a unique silicon chip to perform the basic circuit for equipment using a single process?

The answer to these questions resulted in the integrated circuit or IC. The integrated circuit consists of a tiny silicon chip that houses a determined number of components, such as transistors, resistors, capacitors, and diodes, interconnected by conductive paths that act as wiring. The components and paths are formed by a complex process that includes diffusion, photo printing, and etching.

Figure 171 gives an idea how the many components can be created on a silicon chip.

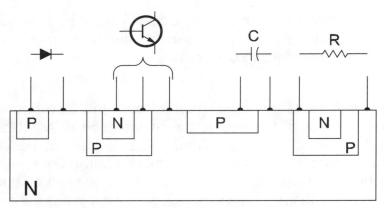

Figure 171

In an N substrate, P regions are created forming a diode. If in the P region of a diode, another N region is formed, a transistor will be created. A capacitor can be formed by a large P region. Having a reverse bias at the junction to create a capacitor, the junction will act as the dielectric

Section 11—Electronic Components

and the PN regions as places similar to the varicaps. A resistor can be formed by a P channel formed in the N region. The width and length of the channel determine the resistance of this resistor. The number of components, their disposition in the chip, and the wiring are planned to perform as a complete circuit such as an amplifier, a voltage regulator, or even a computer.

It is possible to house in a single chip, circuits with few components for simple functions to circuits containing more than 7 million components, such as in modern microprocessors used in personal computers. Even a complex integrated circuit with millions of transistors has a tiny footprint measuring only in millimeters.

The principal advantage of the integrated circuit is that the circuit can be manufactured almost complete, needing only a few complementary support devices to perform in the desired equipment.

For example, it is possible to create an IC containing a complete audio amplifier that needs only the input jacks, the power supply, the volume control, and the loudspeaker to perform the complete equipment. The integration has some limits. Some components exist that can't be easily integrated on a silicon chip. For instance, large-value capacitors and inductors are components that can't be found in integrated circuits. So, when planning a complex integrated circuit, if large-value capacitors are needed, they must be external, and the common traditional discrete type must be used.

Although the integrated circuit is very important as a simplified solution for projects it can't form a complete piece of equipment or circuit in many cases, needing completion with external components as capacitors and inductors.

Since an IC can contain practically any configuration with any number of components, they exist in a very large number of types and functions.

Symbol and Types

Figure 172 shows the common symbols adopted to represent an integrated circuit and its various types.

At the side of the IC's symbol or inside, an identification figure allows the technician to identify it.

Electronics for the Electrician

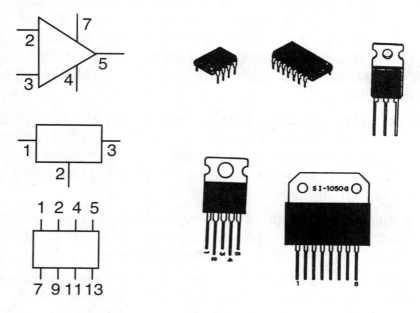

Figure 172

The number of terminals varies according the function. Types with number of terminals ranging from eight to more than 40 are used in common functions. Types with more than 200 terminals are found in special applications such as computers.

Figure 172 shows the most common types of ICs, found in many pieces of electronic equipment that use DIL (Dual In Line) and SIL (Single In Line) cases. See the number of the terminals starting from a mark or point in the case. The identification of the terminals is very important when replacing an IC with another or when trying to troubleshoot. Voltage measurements must be made in these components. Some integrated circuits that perform high-power functions like motor controls, voltage regulators, and audio amplifiers are equipped with tabs or resources to allow them to be on heat sinks.

Specifications and Types

Today we can say that there exist millions of different types of integrated circuits. They are identified by a part number. Only with specific information about the part number is it possible to know what an IC is and what it does. The electrician who wants to work with electronic compo-

Section 11—Electronic Components

nents doesn't need to have complete handbooks or manuals of all the integrated circuit as it is impossible.

Common functions

Some integrated circuits contain a configuration simple enough and useful enough to allow them to be used in a large number of applications. So, these circuits act as basic blocks in many projects and can be found not only in one kind of application, but in a large number of applications. For example, an integrated circuit that contains a simple low-frequency stage with some transistors and some other components can be used in an audio amplifier, a low-frequency oscillator, an alarm, a motor, and many other circuits.

Many types of these general-purpose or common-function integrated circuits were created and many of them became popular and are used in all the applications we can imagine. Electronics magazines that give projects to hobbyists indicate those ICs in their projects as they can be found in any electronic parts shop and can be handled easily.

The common functions found in ICs are divided in two other groups:

Analog

Analog integrated circuits are the ones that work with DC and analog signals, such as audio, video, AC, and others. We define as an analog quantity, one that can assume any value between two limits. An AC voltage is an analog quantity as its amplitude can assume any value between zero and the maximum amplitude. On the other hand, a digital quantity can assume only determined values between two limits. A byte can assume only the integer values between 0000 0000 and 1111 1111.

In the group of analog ICs we can find some functioning as voltage regulators, operational amplifiers, comparators, audio amplifiers, video amplifiers, timers, PLLs, etc. It is interesting to give a brief description of each of these ICs as they are very common in many appliances and will be found by the electrician in electric installations and in many AC-powered appliances.

Voltage regulators

The purpose of a voltage regulator is to maintain constant the voltage in a circuit. The power supplies of many pieces of equipment use this type of IC to have a regulated output supply voltage. Today a three-terminal, 1-A voltage regulator, such as the 78XX and 79XX series or family (where the XX is replaced by a number indicating output voltage), is very common in appliances. For example, the 7806 furnishes a 6 V output under 1 A of current (maximum). Another type of voltage regulator is the one used in SMPS (Switch-Mode Power Supplies) and the ones found in computers, video monitors, and many other modern appliances.

The operation principles of these regulators are different from the linear regulators as they acts as oscillators, producing a voltage that depends of the frequency and pulse width of the output signal.

Symbols and Types

Figure 173 shows the symbols and some types of this IC. Notice that many of the ICs used as voltage regulators have resources to be mounted on a heat sink. Observe also that other ICs can have more than three terminals.

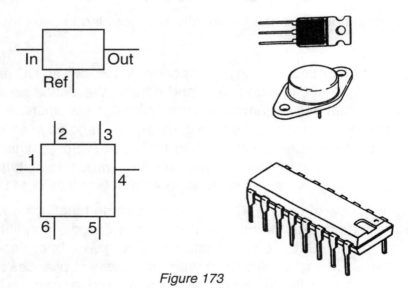

Figure 173

Specifications

The ICs used as voltage regulators, like any other IC, are specified by a part number. The electrical characteristics are found in data sourced by the manufacturer. The main specifications of these kind of IC are:

Section 11—Electronic Components

A. Input voltage range—the range of voltages that can be applied to the input. In the documentation it can be abbreviated by Vin.
B. Output voltage or range (Vout)—Some ICs are fixed voltage regulators. They have an internal zener that fixes the output voltage. Other have resources to be programmed by an external zener or resistors to a determined output voltage. Figure 174 shows a circuit, the LM350, which is an adjustable voltage regulator for a large range of voltages and output currents up to 3 A.

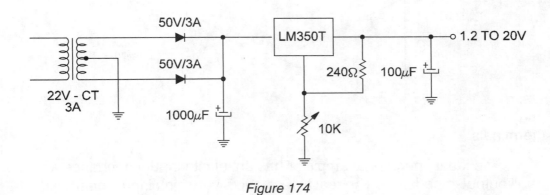

Figure 174

C. Output current—the maximum output current that can be regulated by the IC is an important parameter to be observed when using this kind of IC.

Where they are found

Voltage regulator ICs are present in almost all equipment plugged to the AC power line and even in battery-powered circuits acting as voltage reducers. After rectifying and filtering, they convert the AC power line voltage in one or more DC voltages necessary to power the electronic circuits. These ICs are present in the first stage of any equipment or "power supply" stage.

Testing

The simplest way to test this kind of circuit is measuring the input and the output voltages. The multimeter is the instrument recommended for this task.

Electronics for the Electrician

Operational Amplifiers

Operational Amplifiers (AO) are general-purpose amplifiers originally designed to perform mathematical operations in analog computers. An operational amplifier has a noninverting input represented by a (+) and an inverting input indicated by a (–). The signals applied to the inverting input appear in the output with the phase shifted, as shown in Figure 175.

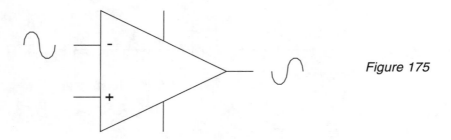

Figure 175

The ideal operational amplifier has an infinite input impedance and a null output impedance. The real types have a very high input impedance of many megohms and a very low output impedance (some ohms to hundreds of ohms).

The gain of an operational amplifier or the number of the times the output voltage is higher than the voltage applied to the input can be programmed by a feedback circuit as shown in Figure 176.

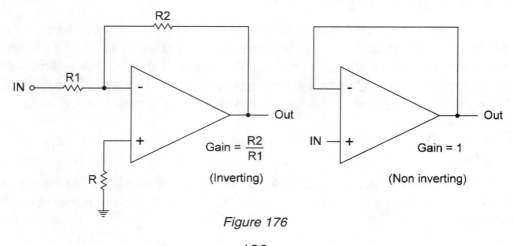

Figure 176

The higher gain is achieved when the feedback is infinite and typically ranges from 10,000 to 1,000,000 in the common types. The lowest gain is achieved when the feedback loop is a zero ohm resistance and is one.

The configuration with a voltage gain of one is called voltage follower. In a voltage follower the output voltage is the same as the input voltage. Despite this, the difference in impedance between the input and output makes the circuit an amplifier with a very high current gain.

Operational amplifiers can be found in a large number of types, sizes, and presenting a large range of electrical characteristics (input and output impedances, power supply voltages, gains, type of transistors found inside as FETs or bipolars, operation frequency range, etc.).

Symbol

Figure 177 shows the symbol adopted to represent an operational amplifier. They generally look the same as other ICs.

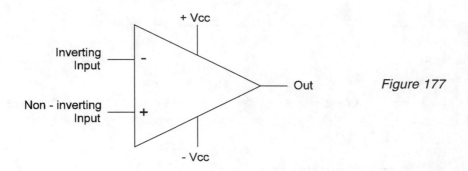

Figure 177

Notice that some ICs exist that contain two or more operational amplifiers. In some cases we can also find terminals where additional components are plugged. These external components are used to add special characteristics to the circuit such as frequency compensation.

Specifications

The operational amplifiers are indicated by a part number according to a code adopted by the manufacturer. Normally a group of letters is used to identify the manufacturer—MC = Motorola, SN = Texas Instruments, LM = National Semiconductor, etc.

Electronics for the Electrician

From the part number or consulting data sheets and/or handbooks it is possible to access the electrical characteristics of each device. These characteristics are:

A. Open loop gain—The gain is determined by the internal circuit and is specified the maximum gain or the "open loop" gain where no feedback is used. The values can range from 1000 to 1,000,000 in common types. The gain of an AO can be abbreviated by G.

B. Power supply voltage range—This specification indicates the range of voltage that the operational amplifier needs for correct operation. It can range from 1.5 V to more than 40 V.

At this point it is interesting to observe that in many applications the AO needs a symmetrical or double power supply. It is a power supply where a positive and a negative voltage referred to the ground (0 V) must be generated as shown in Figure 178.

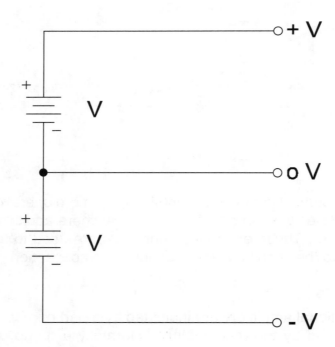

Figure 178

Section II—Electronic Components

In some documents, data sheets, and diagrams it is common to refer the power supply voltage for an AO as 6-0-6 V, +6, -6, 0 V, +6/-6/0 V or 6+6 V if it needs a symmetrical supply as the one shown in Figure 178.

C. Gain x Bandwidth—The gain of an operational amplifier falls at the same time the signal frequency increases. The frequency band is limited by the point where the gain falls to one; this determines the gain bandwidth product. Common operational amplifiers can operate in frequencies from 1 or 2 MHz to up to some hundreds megahertz.

D. CMRR—If two signals with the same amplitude and phase are applied, one to the inverting input and the other to the noninverting input of an operational amplifier they must be cancelled and the output will be zero. In practice, due to small differences in the internal components, this is not possible and some part of the signal is not cancelled, which then appears in the output. How efficient is the AO when canceling a signal in this operation mode is given by a specification called Common Mode Rejection Ratio or CMRR. This specification must be as high as possible and is indicated in decibels (dB). Common types have CMMR as high as 90 dB and more.

Where they are found

Many appliances use operational amplifiers in the integrated circuit form. They normally are common types such as the 741, CA3140, TL080, TL081, and many others that are easily found in any electronic components shop. Special operational amplifiers can be found driving step motors or other devices, but in general, they can be found at a dealer by part numbers, too.

The basic function of an operational amplifier in many circuits is the amplification of signals coming from sensors, transducers, or other sources, driving them to powerful output stages or circuits that must work with signals having a larger amplitude.

Testing

In many applications the operational amplifier can be tested measuring the voltages in their pins. If the voltages are altered, and the components plugged into it are good, the problem will be certainly in the OA. Another way to test is using a special probe circuit.

Audio Amplifiers

Complete audio amplifiers with output powers ranging from some milliwatts to more than 100 watts can be found in the form of a linear integrated circuit. These integrated circuits consist, in some stages of amplification, of the preamplifier, driver, and output stages interconnected to perform as an audio amplifier.

The high-value capacitors necessary to many of the stages are connected to the circuit by external pins. In many cases, the manufacturers can use the circuit with a minimum of external capacitors.

Many audio amplifiers can have resources to program the gain or the frequency response by some external components. In the simplest types we have an input where the volume control is placed, an output where loudspeaker or headphones/earphones are connected, the power supply pins, and some other necessary parts, such as capacitors (coupling and decoupling) and resistors (to program gain). Many manufacturers have types containing two amplifiers in one case, making the project of stereo equipment easier.

Symbol and Example

Figure 179 shows the basic symbol adopted to represent an audio amplifier. The cases are SIL or DIL or another type based on the output power.

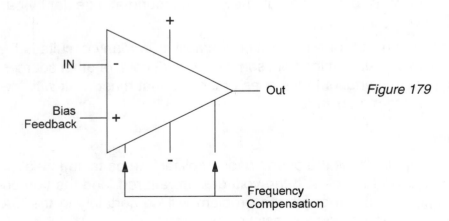

Figure 179

Specifications

The ICs of this group are also identified by a part number according to the manufacturer. The electric specifications of an audio amplifier in the form of an integrated circuit are the same for any other audio amplifier.

A. Output power—the amount of power that the circuit can drive to a loudspeaker. It is indicated in watts (W) and in some cases abbreviated by Po or Pout.

B. Power supply voltages and current—important when designing the power supply for the circuit. It is given in volts.

C. Output impedance—characteristics of the loudspeakers/earphones connected to the output of any audio amplifier are very important to not overload. Impedance is given in ohms and typically is between 2 and 16 ohms for types designed to drive loudspeakers and higher for earphone/headphones amplifiers.

D. Input impedance—given in ohms.

E. Gain—how many times the amplitude in volts of the output signal is higher than the amplitude of the input signal. This specification is important for the input impedance to determine the sensitivity of an amplifier.

F. Operation curves and other specifications—performance of an audio amplifier according to the signal frequency is given by curves. These response curves show how the fidelity of an amplifier is when working with signals of the audio spectrum. These specifications are important when making a project. For the technician who needs information to find a replacement, the part number is enough. It is important to call the reader's attention to the fact that amplifiers with the same power in the cases can be different in disposition of pins and the external components meaning that they are not interchangeable.

Where they are found

Audio amplifiers are found in every part and a large quantity of common AC-powered appliances, embedded in electric installations, or even powered from batteries, but all related to applications in the electric installation. For instance, intercoms, alarms, doorbells, wireless telephones, wireless intercoms, baby monitors, and ambient sound sources can use an audio amplifier (or several) in the form of an integrated circuit to drive an earphone or one or more loudspeakers.

When creating the equipment the designer must instead use a dedicated integrated circuit where all the functions are embedded in a unique device and can use some common integrated circuits in some functions as the audio output.

Testing

The best way to test a common integrated circuit, such as an audio amplifier, is by measuring the voltages in their pins. It is recommended to do this measurement only after verifying if the associated components (external capacitors, resistors, etc.) are good. For this, it is important to have an idea about the magnitude of voltages to be found or to have the schematic diagram of the equipment where the voltages are indicated.

Timers

Another important group of the linear integrated circuits family is the one formed by timers. These are circuits that can be used to generate a signal or to count a time interval. Basically, they are circuits that can be used in two operation modes: monostable and bistable.

In the monostable mode, once triggered, the circuit produces a signal during or after a time interval triggering something or controlling something as suggested in Figure 180.

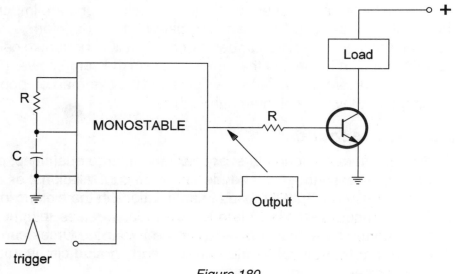

Figure 180

Section II—Electronic Components

In the unstable mode, the circuit generates a square signal with a frequency dependent on some external components, normally an R-C network as shown in Figure 181.

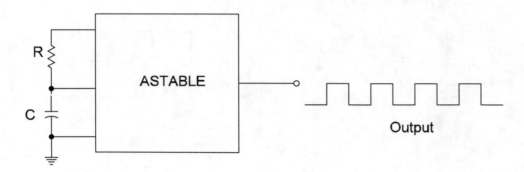

Figure 181

Symbol

Figure 182 shows the symbol used to represent this function. The ICs of this family are found in DIL cases and others.

The 555

The most common of the timers is the 555 that can be found with manufacturer designations such as NE555, TLC7555, LM555, and MC555.

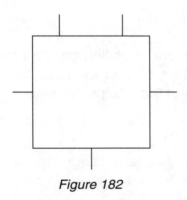

Figure 182

This integrated circuit can operate in two modes: unstable and monostable depending only on the way some external components are connected as shown in Figure 183.

The bipolar version (where the internal transistors are bipolar types) can provide time intervals up to one hour and oscillate in frequencies up to 500 kHz. A version using JFET transistors (TLC7555) offers larger time intervals and produces higher frequencies with lower current consumption. The meaning of 555 to electronics can be evaluated by a recent report that said that more than 1 billion units of this device type were sold since it launched.

Electronics for the Electrician

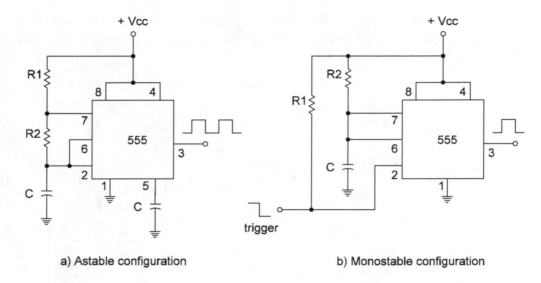

a) Astable configuration   b) Monostable configuration

Figure 183

Specifications

The timers in the form of integrated circuits are specified by a part number just as others. The electrical characteristics can be found in the manufacturer's documentation or the schematics of the equipment where they are being used.

The main electrical specifications are:

A. Power supply voltage—the given power supply voltage range where the circuit can operate.
B. How to calculate the time interval or frequency—In the applications, the external components (R and C) determine the operation frequency. A formula or curve allows the designer to find the correct value for the components to any application.
C. Output current—This is an important piece of information, as it says what can be controlled from the IC and how it can be done. In some cases, a transistor stage must be used to control something with a large current consumption .
D. Operation modes—Many timers have different operation modes. They are given in the specifications with information about the way the external component must be connected to perform.

Section II—Electronic Components

Where they are found

There are many small devices found in electric installations that are based in timers like the 555, TLC7555, and many others. When you press a pushbutton in the wall and the lamp stays on for a certain time interval, most likely the circuit controlling this includes a timer integrated circuit like the 555 or one of the same family.

Testing

As in many other integrated circuits, there isn't a special procedure to test them. Any integrated circuit is a different circuit and specific procedures should be developed to make an effective test. The best way to know if the IC is good is by reading the voltages at its pins. Start with the idea that all other components are good, and compare voltages with specifications or indications in a diagram.

Phase Locked Loop—PLL

Phase Locked Loop, or PLL, are circuits that can be used in many important applications related to electric installation, home appliances, and cars. A PLL is a circuit that can recognize a signal of a determined frequency acting on a relay or other device. If the signal is modulated in frequency, the PLL can also be used to extract the modulation signal. Many types of PLLs in the integrated circuit form are found.

Symbol

Figure 184 shows the symbol of a PLL. The body is not different from any other linear function described here. The only way to know if the IC is a PLL is by its part number displayed as a number on the component's body.

Specifications

As any other IC, the PLLs are indicated by a part number. Using the part number, it is important to know some electric specifications when using a PLL or when trying to find a replacement type. The main specifications are:

A. Power supply voltage range—the voltage range in which the device functions.

B. Frequency range—a very important specification as it indicates the maximum frequency of a signal that can be recognized by a PLL or generated by it when operating as oscillator. Common general purpose PLLs, such as the 556, can operate at frequencies up to 500 kHz.
C. Sensitivity—the minimum amplitude of a signal that can be recognized by a PLL.
D. Output current—this speci-fication indicates the amount of current the device can source or drain when driving a load.

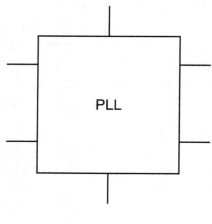

Figure 184

Where they are found

In many modern applications, the PLLs are included in more complex ICs. But, some exist that are for a simple application made by small manufacturers, or even made by amateurs that use the PLL as a single chip. In these case some common types, like the 4046, LM567, and others, can be found.

The next applications can give to the reader of how useful are the PLLs.

A. Remote control—used to recognize a tone sent by the transmitter (using a radio signal or IR) and act on a device (the motor of a garage door, or a function of an electric or electronic appliance)
B. Intercoms—Some types use the AC power line to send the sounds. The sound picked up by a microphone is used to modulate a circuit producing a frequency-modulated signal that is applied to the AC power line. A PLL in the receiver detects this signal and applies it to an amplifier.
C. Alarms—a tone is produced by a circuit and applied to a PLL via sensors. If any sensor is activated the PLL detects the absence of the tone and triggers the alarm.

Section II—Electronic Components

Testing

The same procedure as recommended for others ICs is valid here: measure the voltage at the pins and compare it with the expected values.

Other Linear Functions

Many other functions can be found in linear ICs. Besides those described before, the reader can find ICs containing video amplifiers, comparators, voltage references, radio receivers, etc. There isn't a specific symbol for these functions and their aspects are the same found in other ICs.

Digital

The digital integrated circuits form a special group of devices with a large range of applications in modern electronics. The basic idea comes from digital electronics. Some fundamentals of digital electronics are important to know to expand information about the devices of this important group. Digital electronics start from the idea that any quantity can be represented by zeroes and ones (0 and 1).

Using only two digits to represent any quantity results in a numeration system named binary. Figure 185 shows how a number in the decimal system can be converted to the binary system.

The advantage of the use of binary representation of any quantity is that it can more easily be handled by electronic circuits. In the simplest form, an open switch can be used to represent a zero and a closed switch can be used to represent a 1 as shown in Figure 186.

Working with opened and closed switches it is possible to make any mathematical or logic operations with numbers. The rules that command these operations were first established by Robert Boole two centuries ago. Moving now to the idea that a transistor (and some other components),

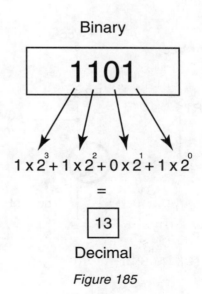

Binary

$$1101$$

$$1 \times 2^3 + 1 \times 2^2 + 0 \times 2^1 + 1 \times 2^0$$

$$=$$

$$13$$

Decimal

Figure 185

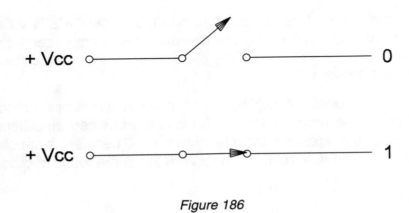

Figure 186

also act as a switch, it can be used not only to represent a zero or a 1 but to execute operations with binary numbers.

As shown in Figure 187, if a transistor is switched to a saturated state (conducting) the voltage in its collector falls to zero or a value near zero, allowing it to represent a low logic level or zero.

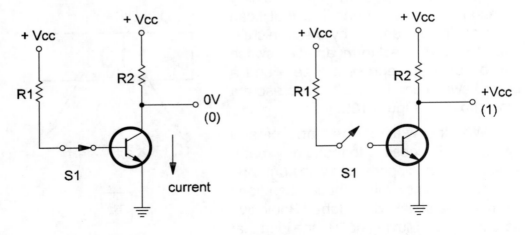

Figure 187

Section 11—Electronic Components

On the other hand, if the transistor is switched off (cut), the voltage in its collector rises to a value near the power supply value. This voltage can be used to represent a high logic level or a 1.

Using many parallel transistors any complete number can be represented, and commanding them with appropriate signals, we can execute mathematics or logic operations.

A single digit, 1 or 0, that is used to represent a binary number, is called a bit and a group of eight bits is called a byte. A group of four bits is also called a nibble. A byte can be used to represent any quantity or decimal number between 0 (0000 0000) and 255 (1111 1111).

Thousands, and even millions, of transistors wired as switches can be placed in a chip in configurations that can perform any operation with binary numbers processing the bits to perform final bytes. They are digital circuits like the ones found in computers, microprocessors, microcontrollers, etc.

A certain number of basic functions are used in the projects of digital circuits. These functions are represented in Figure 188.

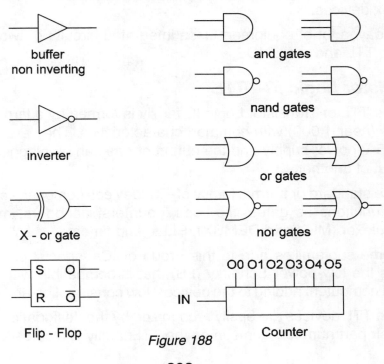

Figure 188

In the first group of basic functions are the gates. Gates are circuits that combine two or more logic levels or digital signals given as a result of logic level or signals determined by rules. The rules follow the mathematics created by Boole. In the second group are more complex functions such as flip-flops (that can store a bit), multiplexers, counters, and decoders. The readers who want to know how a computer and digital circuits work must study digital electronics and Boolean algebra.

In the first computers, like the ENIAC (1942), operations with the bits were done by circuits using tubes. Some time later, the tubes were replaced by transistors. The next step was the use of a large quantity of integrated circuits, each containing many transistors, placed in boards. Now almost all the circuits are inside a chip (one integrated circuit) forming processing units called microprocessors.

Although all the digital circuits are now inside the chip of a microcomputer or other advanced digital circuit, there are applications where integrated circuits performing simple operations or functions must be used. These circuits form families of digital integrated circuits were they all have matched characteristics allowing for interconnection directly to form more complex devices.

Today, a lot of equipment uses integrated circuits of two main families, the TTL and the CMOS.

Transistor Logic IC—TTL

The TTL or Transistor Logic IC family is formed by a large number of devices (near 1,000) with common characteristics. They are all powered from a 5 V power supply and the output of one can be plugged directly to the input of another.

The standard or normal family of TTL devices comprises a large number of functions as gates, flip-flops, counters, decoders, multiplexers, demultiplexer (MUX and DEMUX), PLLs, and timers.

Some subfamilies add to this group of ICs' special characteristics, such as the low-power Schottky (LS) that is compatible with the circuits found in computers adding to the devices' low consumption and high speed.

The TTL devices are easily recognized by the designation "74" beginning their part number. Sometimes the TTL family is also called 74 family

Section II—Electronic Components

or 74xx. Almost all devices in this family are designated by a 74 followed by a number according to their function (7400, 7492, 7191, etc.). The subfamilies are recognized by a letter or two after the 74. For example: 74LS04 (low-power Schottky), 74H93 (high power), 74S121 (Schottky), etc.

CMOS

The other important logic family of ICs is the one formed by the CMOS integrated circuits. These circuits perform the same functions found in the TTL but with different electric characteristics. They can be powered from supply voltages ranging from 3 to 15 volts or 3 to 18 volts. They are not as fast as the TTL, but their consumption is lower and their input impedance is very high.

The devices of this family are designated by a "40" followed by a number that says what the device does. So, we find designations like 4004, 4017, 4093, and 40121. Some CMOS subfamilies use the same designation or number of the equivalent TTL function. These devices are indicated by the 74 code but with a C to indicate it is a CMOS (i.e., 74C00).

When working with TTL or CMOS circuits, the technician needs a digital CMOS or TTL handbook or databook to reference what the function of the IC is. The important fact to be observed when working with logic circuits is that all of them work with pulses or logic levels. If the circuit operates with logic levels in the nonsynchronized mode, in the inputs and outputs you will find only two possible voltages: 0 V or the power supply voltage (5 V in TTL or between 3 and 15 V in CMOS circuits), as shown in Figure 189.

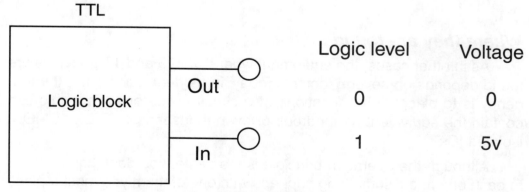

Figure 189

If the circuit is synchronized by a clock, the signals are square pulses as shown in Figure 190.

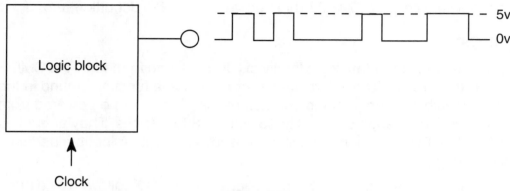

Figure 190

Specifications

The function of each digital integrated circuit is given by its part number. The basic electric characteristics are the same in the entire family. It is important to remember that TTLs always need 5 V to be powered and CMOS needs any voltage between 3 and 15 V.

> **WARNING**—The integrated circuits of the CMOS family are very sensitive to electric charges stored in the body (static electricity) as are the CMOS transistors. Never touch the terminals of these devices directly. The discharge of the electric charge stored in your body can destroy the device.

Where they are found

As in other cases, the equipment where CMOS and TTL devices are found depends on several factors. Today, in modern equipment, the tendency is to place all the functions of a circuit inside one. This chip can contain the equivalent of hundreds or even thousands of TTL or CMOS-isolated ICs.

Although the operation principle is the same, you don't have access to each one and if something happens with one of them you must replace

Section II—Electronic Components

all of the circuits or the IC. On the other hand, the electrician, in his work with electronics, can find older equipment or equipment made by small manufacturers that doesn't use complex circuits, and also where isolated TTL or CMOS circuits are found.

Their circuits perform simple functions, such as those of timers, sequencers (light effects and signaling), alarms, remote controls, automatic lamps, night lamps, and inverters.

A basic circuit can show to the reader how they are used, such as the interesting circuit found in common applications that is shown in Figure 191.

This is a sequencer circuit using the common 4017. Each pulse you apply to the input makes one output of the circuit go to the high logic level, which means the voltage rises from zero to the power supply voltage. It is possible to select which output goes to the high logic level or is activated by the number of pulses applied to the input.

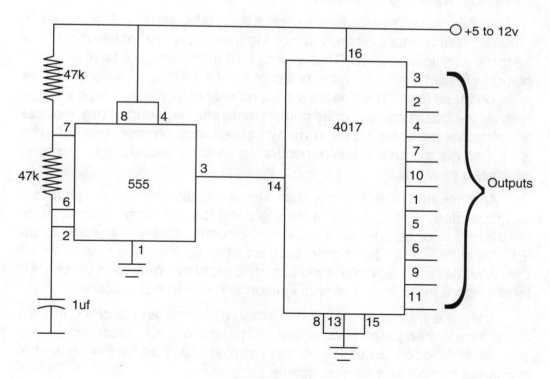

Figure 191

Testing

Testing of digital integrated circuits must be done using special instruments. The simplest way is measuring the voltages with a multimeter, but it is not conclusive in many cases (depending the way the circuit operates). The advanced test uses logic probes or oscilloscopes to monitor the digital signals in each IC.

Custom and Embedded

Today almost any function or circuit can be made in the form of an integrated circuit. To reduce costs and space the manufacturers can create their applications and put almost all the circuits in a single chip. This means that their products don't use many common integrated circuits, but instead use only one circuit dedicated to a specific use. In some equipment using these integrated circuits, functions like the power output can be performed by common transistors or SCRs, but the core of the circuit is a single chip with no equivalents.

As the cost of ICs for these applications falls, along with the quantity of equipment manufactured, the boards these ICs are installed on have become very inexpensive. This means that in the case of failures, entire boards are replaced. The main problem for the technician who makes repairs on these is that the boards are found only in authorized repair shops. Neither the board nor any of its components can be found at regular dealers. Another problem is that in many cases the equipment is a one-way type. This means once a failure occurs it is better to discard the equipment and buy a new unit.

Another kind of technology found in new equipment is one referred to as embedded. Embedded devices are used to control, monitor, or assist the operation of equipment, machinery, or entire areas, according to the IEE definition. These components are an integral part of the system where they operate. In a car, embedded circuits control all operations of the electrical system including electronic ignition and electronic injection.

They are basically formed by microprocessors and microcontrollers in the form of integrated circuits that are mounted inside black boxes. Access to embedded circuits is through cables. They allow the system to exchange control and sensor signals.

Section II—Electronic Components

Figure 192

Types

No special symbol has been adopted to represent this kind of circuit. Figure 192 shows various types of these devices.

As the device is not intended to be handled by technicians, but only by machines, it is common to find them on boards in the SMD format, or a format called on-board. In the on-board format the chip, without case, is soldered directly on the printed circuit board and protected by a rosin or epoxy.

Specifications

In some cases the manufacturer indicates the IC by a part number, but in some one-way applications the device could only be identified by the manufacturer. In some cases, as in the black boxes of many embedded systems, the devices do not have any identification displayed.

Where they are found

The embedded functions are found in car equipment, home appliances, and toys just to name a few. For example, you can find these ICs in calculators where the manufacturer "creates" the circuit he needs to perform a function and places it directly on the board. Another example is electronic dolls that have a chip with voice-programmed functions so it can talk. The low cost of these specialized circuits and the form in which they are mounted make them impossible to be repaired or replaced.

Clocks and watches also use special chips containing the complete function they need in a single chip. Some of them can include the display. The clocks, counters, and other functions formed by a chip and the display

Electronics for the Electrician

are called modules. Many appliances use complete modules of counters, meters, clocks, and other basic functions. In some cases, if the function is not critical, the module can be replaced by another that has the same function.

Testing

There is no practical way to test them as each represents a separate function.

Microprocessors and Microcontrollers

Microprocessors and microcontrollers represent another great group of devices in the IC family. In a single chip thousands and even millions of components are used to perform a complex digital/analog function. Microcontrollers are intended to control applications, such as those in industrial machines and automatic devices. The microcontroller is formed by input/output blocks, a processing unit, and a task-programmed memory.

Figure 193 shows the structure of a typical microcontroller block.

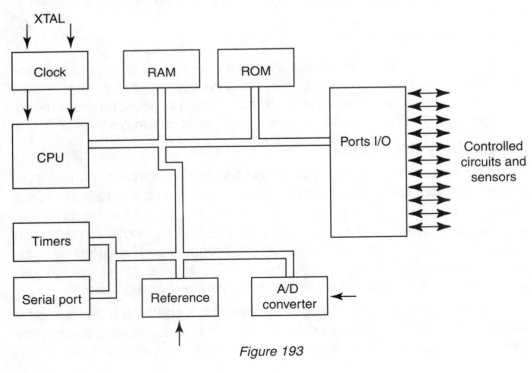

Figure 193

Section II—Electronic Components

Microprocessors are also used to perform complex mathematical and logic operations and are used in computers and other equipment. The microprocessor only works with digital data performing operations and returning with the results in the digital form. Figure 194 shows the block diagram of a typical microprocessor.

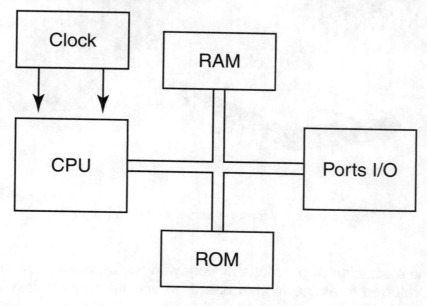

Figure 194

Microprocessors have input and output ports (I/O) where the information is placed and gathered, a CPU (central processing unit) where the mathematical and logic calculations are made, a memory, and other supporting circuits. Microprocessors are found in computers. Some families of microprocessors exist that are very common today.

Starting with the Z80, an 8080 is found in many old appliances as a "brain" where functions are calculated. More often now, we are using the microprocessors of the X86 family. The "X" is a number indicating the specific type. Several years ago, the 286 was common, now it is the 586. Variations with different names that include the Pentium K6, Athlon, and Celeron are prevalent. These microprocessors become famous through recognition in the personal computer (PC) field where they are used as the core.

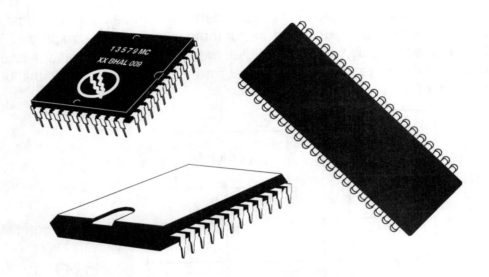

*Figure 195
Chips and microprocessors similar to those used in PCs*

The electrician who intends to work with computers will find a board inside, referred to as the mother board, where the microprocessor (also called CPU or central processing unit) is placed. The microprocessor is a very complex integrated circuit, containing millions of transistors and using a special body as shown in Figure 195.

Microcontrollers are found in industrial machines, appliances, cars, and other equipment performing control functions. They appear in the form of common integrated circuits in DIL bodies or more complex circuits (SMD and other) based on the number of pins and functions. Some families of microcontrollers are popular and are used in many applications. The 80C51, PIC, COP8, and others can be found in professional equipment.

The use of a microcontroller depends on some resources. The designer must have the tools to program them including a special board with a socket where the microcontroller is placed, a computer, and software to prepare the program. The designer should prepare the program that says what the microcontroller must do. Once prepared, the PC transfers the program to the microcontroller.

Section II—Electronic Components

Specifications

Microcontrollers and microprocessors have large databooks to provide all their specifications. The technician wanting to use these devices needs special courses that involve a large amount of technical information.

Digital Signal Processor—DSP

Many other devices in the form of ICs can be found in electronic equipment including those related to electric installations and cars. New functions like the DSP are an example of devices that are designed to perform specific functions normally determined by the manufacturer (custom).

These devices are found in many kinds of electronic equipment, such as cellular telephones, computers, fax machines, industrial machine controls, dishwashers, clothes washers, remote controls, electronic injection, and ignition controls in cars, DVD players, TV satellite receivers, and videocassette players. They are not simple devices or components, but complete circuits programmed for a predetermined purpose, mounted on a chip, and installed inside a case like the ones used in any other integrated circuit.

What is a DSP?

The traditional analog circuit operates with analog signals like sounds, images, or voltage changes from a sensor, along with traditional components and circuits such as transistors and filters. The analog circuit uses the electric properties of the components to introduce some alterations in the signal's original form. For example, a bandpass filter lets only signals for a determined frequency band pass through it; an amplifier changes the amplitude of a signal.

Depending on what you want to do with a signal, the implementation of a network using common devices is not only very difficult, but is not very accurate. However, if we consider that any analog signal can be represented by number or a stream or numbers (converted to a digital form), why not perform the desired function with the signal working in the form of numbers? Instead of using a network that lets only a determined frequency band pass, convert the signal into a stream of numbers in a computer, apply an algorithm to the numbers that recognizes the frequencies and amplitudes, and the undesirables values are cut. (An algorithm is a sequence of mathematical and logic procedures.) After that, the remaining values are converted again to the analog form.

Figure 196 shows, by a block diagram, how the DSP works.

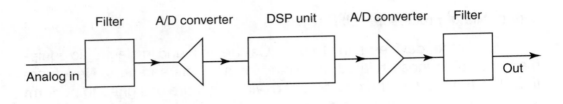

Figure 196

Since a computer is precise and fast, the results presented by this kind of circuit are fantastic—much better than a common analog circuit designed to do the same task. The processor inside a DSP is fast and can perform extremely complex operations, making it possible to make almost everything you want with a signal depending only on the software (algorithm) used. Using a DSP, what a circuit will do no longer depends only on the components used, but on a program!

Many DSPs are used today in common electric and electronic appliances in control functions to replace microcontrollers. For example, a manufacturer can buy an empty DSP and program it to perform all the functions needed to control the motor of a clothes washer, including functioning as the water-level sensor and the temperature sensor. The DSP processes this information and determines the voltage (speed) to be applied to the motor, the voltage applied to the water heater, etc.

To install it on a board, there are specific DSPs for this product without the need of common devices in a large board. The same DSP can be used by another manufacturer to control the motors, solenoid, and relays of an industrial machine. The only thing the other manufacturer has to do is plug the necessary peripherals into the DSP and install the appropriate software.

Of course, if a DSP fails in a machine, the technician has no other solution than to look for the original part from the manufacturer. Also, a single DSP (without the program) doesn't work in a machine if it isn't pro-

Section II—Electronic Components

grammed with the original software. The DSP must be purchased with the original programming of the manufacturer. The problem is simpler if a peripheral component fails, such as a transistor, SCR, or relay. In this case, the component can be easily replaced.

It is important to know that the DSP is identified by a part number (Texas Instruments, Analog Devices, and Motorola are manufacturers of DSPs), but if you buy one at any dealer it comes without the programming. If you intend to build any electronic equipment, depending on its function, the use of a DSP should be considered.

Surface-Mounted Devices—SMDs

Many of the devices we have seen before can be found in an ultraminiaturized version mounted by machines in large manufacturers' assembly lines. Notice the majority of the devices are in cases. The active element of a resistor is only a thin layer of carbon; the active element of a transistor, SCR, or triac is a small silicon chip. The active element of an integrated circuit is also only a small silicon chip.

The body, for the most part, represents more than 95 percent of the size and weight of the component. It is clear that, if the body is reduced, we can gain space and profit, but on the other hand, we lose the ability to handle it. The basic idea of the SMD is to reduce the body to the minimum,

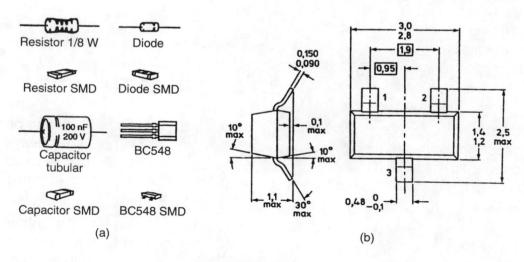

Figure 197

in a way that only machines can handle it. Reducing the size of the components in the equipment makes them cheaper, smaller, and lighter. Resistors, capacitors, diodes, transistors, and ICs can all be found in SMD versions as shown in Figure 197.

This figure shows the typical dimensions of these components and how they are mounted on a printed circuit board. Special machines place the components on a board using glue and solder. The working technology applied in this case is also called SMT (surface-mounting technology), so it is common to refer to devices used in this form as SMD and the technology as SMT.

It is important for the electrician to understand that the functions and the circuits are the same as those used in the common components. The difference is only in the size and shape of the components. The main problem is how to handle them if a repair is needed.

Special kits, consisting of small tweezers, screwdrivers, and other delicate tools are available to help technicians replace SMD devices when repairing equipment. The same dealers usually can furnish resistors, capacitors, and many other components in the SMD format as well.

Section 3

Troubleshooting and Repair

One of the purposes of this book is to give an electrician some basic knowledge about electronic components and circuits—knowledge that is necessary to make simple repairs and install electronic equipment coupled to electric installations, powered from electric installations, or even found in a car. It is not up to the electrician or professional of any field other than electronics to make complex repairs in electronic equipment. This task must be left to professionals. But, depending on the problem, repairing electronic equipment is not difficult if an electrician knows some basic procedures. There are two kinds of problems with electronic equipment the reader of this book can solve:

1. Simple problems with the equipment circuit.
2. Problems due to the installation or operation of the equipment.

This section presents a basic look at symptoms and causes of simple problems the reader, who may not yet be familiar with the operation of electronic circuits, can solve.

Because we can't see, hear, or smell electricity, knowing exactly what is wrong in a circuit only can be discovered with the aid of instruments or a procedure using symptoms as a basis for troubleshooting.

Of course, as this book is introductory, a reader wanting to go a step further will need more information. If you want to go into the profession, you will have to learn much more in a technical course as well as obtaining access to a variety of technical databases. It is also important and helpful to read technical periodicals because they provide useful procedures and ideas on new technologies.

The First Step

If a piece of electronic equipment doesn't work as expected what are the basic procedures to find the cause and how then to repair it? It is a false idea (coming from the tube age) that you should replace a given part when a certain symptom occurs. This is not valid for very complex equipment and circuits where hundreds or thousands of interdependent components exist. In complex circuits the same symptom can occur when any one of many components fails, and in some cases, many of them fail at the same time. You must know how the circuit works to form a hypothesis about what the cause of the trouble is.

For an electrician who is now starting with electronics, it is too early to try real troubleshooting and repair work in complex equipment. We can, however, provide some basic procedures to allow determination if the cause of a problem is inside the equipment (and requiring professional help), in the installation itself, the result of improper use, or some other external problem.

The basic procedures are also useful to determine the cause of simple problems that need no professional help to be solved, such as burned-out fuses, unplugged connectors, and broken wires.

Safety Rules of Troubleshooting

These are the most important items when working with any piece of AC-powered equipment, and even equipment powered by other sources

Section 3—Troubleshooting and Repair

can use high voltages. It is very important to know the hazards associated with the equipment you are troubleshooting. Electricity can kill and the electrician working with dangerous AC voltages must know that.

- Don't turn on the equipment without thinking about it first. Start with some analytical thinking. Be sure that you know how to operate the equipment you are troubleshooting before powering it on.
- Don't touch parts if you don't know what they are for or how they work. You can cause more damage to the equipment; a simple problem can be turned in an expensive problem.
- Look for visibly damaged parts such as burned-out resistors, interrupted cables, and bad connections. Many problems have simple solutions. The visual inspection is the starting point in any troubleshooting work.
- Learn from mistakes; incorrect procedures when troubleshooting can turn a simple problem into an expensive one. Let a tool fall inside equipment being tested causing a short and you will have a real example of what that means.
- Working with modular parts: modern electronic equipment often uses modular parts. If you can identify the cause of a problem in a module, it is easier to replace the complete module than trying to repair it by looking for burned-out components.
- Don't always trust your instruments. If you are using a multimeter to measure voltages or a resistance in a circuit and the reading is confusing, don't assume that is something wrong with the circuit. In many cases, the characteristic of the multimeter and the presence of other components near the one being tested can mask the results. Using the multimeter is an art and only with experience can you begin to trust all the measurement made by this kind of instrument.
- However, the multimeter is probably the most useful of all the instruments the electronics professional can have. It is very important for someone wanting to become an electronics professional to be familiar with all the uses of this instrument. The next section discusses some more of its uses.
- Film canisters, plastic ice cube trays, eggs trays, and pill bottles can be used for sorting and storing screws and other small parts when you disassemble equipment. Use a notebook to mark the position of any

screw in equipment tested if you have difficulty memorizing it. A leftover screw when reassembling a device can cause a big headache.
- Don't force any part of the enclosure in equipment when disassembling it. If you have to force it, it may be because you are not doing things correctly. The direction you are moving the part is not the correct way or there are more screws to be taken out.
- ESD (Electrostatic Discharge)—Some components, including the CMOS transistors and ICs, are very vulnerable to ESD. Don't touch their terminals when working with them.
- If possible, use a schematic. Many pieces of equipment have manuals or schematics that provide important information for the reader when troubleshooting. If you intend to go further in this field, knowing how to read a schematic is essential to the repair of any equipment.

Schematics

Sams Technical Publishing has circuit diagrams, schematics, or service information for many pieces of electronic equipment sold since the 1940s. You can access the company's service using this Internet URL: *http://www.samswebsite.com.*

You can also reach it at the following address and phone number:

Sams Technical Publishing
5436 W. 78th St.
Indianapolis, Indiana 46268
Customer Service: 1-800-428-7267

Many large public libraries may subscribe to Sams Photofact and have the documents available for photocopying.

Quick Tips for Troubleshooting and Repair

- A totally dead piece of equipment or one with many functions affected may have a defective power supply. Check whether the circuit is correctly powered. For example, see if there is voltage in the AC power line.
- Erratic or intermittent problems are almost always due to bad connections, such as cold solder joints or connectors that need to be cleaned.

- Gradually changing problems (problems that decrease or disappear completely when the equipment warms up) are often due to dried electrolytic capacitors.
- The majority of problems in equipment with many mechanical parts, such as VCRs, CD players, or DVDs, are due to mechanical or optical failures.

The Multimeter

The multimeter, VOM (Volt-Ohm-Milliamp meter), or Multitest is the most important instrument for an electrician who wants to work with electronics. Any test in electronic circuits or electronic components begins with the principle that when it is working, some measurable electric quantity exists. Here, three quantities are the most important: voltage, current, and resistance.

Measuring the current or the voltage in a circuit allows us to determine if the circuit is good or bad. In some cases, measurements can also determine the cause of a problem. Many electronic devices can be tested with some simple resistance measurements.

Analog Multimeters

There are two types of multimeters an electrician can purchase to work on not only electric installations, but also on analog and digital electronics. The analog multimeter is formed by a moving coil meter (microamperimeter) and circuitry with functions selected according the quantities to be measured.

For example, if an electrician must measure resistances he needs to select one of the resistance positions on the switch according to the expected value. If the position ohms x100 is selected, the values read in the scale must be multiplied

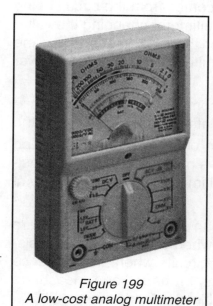

Figure 199
A low-cost analog multimeter

Electronics for the Electrician

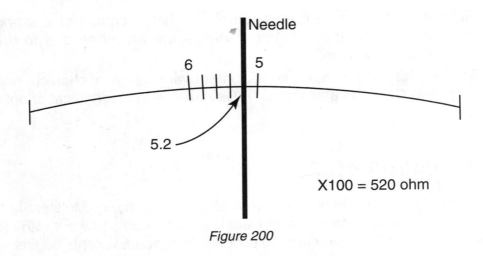

Figure 200

by 100. An indication of 5.2 as shown in Figure 200 indicates a resistance of 520 ohms.

The selection of the quantity to be measured is made by a switch or selected by jacks where the probes are connected. Figure 199 shows this kind of multimeter. The most important specification of an analog multimeter is its sensitivity. The analog multimeter, when measuring voltages, needs power to move its needle. The power comes from the circuit being measured. This power alters or "loads" the multimeter changing the value reading. The better the multimeter, the lower the load needed from the circuit to be measured, meaning a better multimeter can pick up lower voltages.

The sensitivity of an analog multimeter comes from its internal resistance when measuring a voltage (it is associated to the sensitivity of the moving coil meter). This value is indicated in ohms per volt (Ω/V) and should be as high as possible. Less expensive multimeters have a sensitivity of around 1000 ohms per volt. Better multimeters have a sensitivity of 10,000 ohms per volt or more.

When using a low-sensitivity multimeter, an electrician must take care to not be tricked by the voltages indicated when measuring high-resistance or high-impedance circuits. These are circuits where the presence of the instrument can cause great alterations in the voltage measured. This occurs because the presence of the multimeter loads the circuit .

Section 3—Troubleshooting and Repair

Digital Multimeters

Many low-cost digital multimeters are available today. Although by many aspects they are better than the analog type, some functions or measurements exist where the analog multimeter is the better choice. For the electrician who is not familiar with multimeters, using this type of tool is not recommended as a starting point. When beginning, it is easier to use the analog type.

The digital multimeter is formed by an electronic circuit that presents the results of the measurements in a liquid crystal display (LCD) like the one shown in Figure 201. The functions for measurement using a digital multimeter are the same as analog multimeters.

Figure 201
A low-cost digital multimeter

Because the digital multimeter uses an internal power supply (normally a 9V battery), it doesn't need to extract large amounts of energy from the circuit where the voltages must be measured as the analog multimeter does. This means that the circuit being tested interprets it as a very high-impedance load not causing significant changes in the quantity to be measured. The problem of inaccuracy found in analog multimeters when measuring low voltages in high-resistance circuits doesn't exist in the digital types.

How to Use a Multimeter

Voltage measurements.

Electricians must take care not to overload the instrument when measuring voltages. This can cause damage including burning it. They must also differentiate the points where AC and DC voltages are found.

When measuring a DC voltage, place the function selection switch in the scale that gives you a comfortable reading of the expected voltage. For example, if you are expecting a reading of about 12V in a circuit, place the switch in the position 0-15VDC. The position 0-150VDC is not recommended because with 12V, you would have a small deflection of the needle.

Observing the poles, place the probes in the circuit to be measured. If you invert the probes or the poles are different from what is expected, the needle tends to move in the wrong direction, indicating something less than zero. If this happens, invert the probes to correct the reading. Some instruments have a pole switch that inverts the probe connections without moving them. Figure 202 shows how the multimeter is used to measure a DC voltage in a circuit. This process is for digital and analog meters.

Some digital multimeters indicate the value with a "–" sign if the voltage is negative or if the probes are inverted. If you have no idea about the voltage present in a circuit, begin by using a higher scale.

Current measurements

Current measurements are not very common when working with electronic circuits because the circuit where the current is measured must be interrupted to place the multimeter's probes (Figure 202).

When measuring DC currents, polarity must be observed when positioning the probes. It is very important to choose the correct scale when measuring currents. If a low-current scale is chosen in a high-current circuit, the meter can be damaged or the internal fuses can burn. Use the higher current if you have no idea about the value of a current to be measured in a circuit. The procedure is the same whether using a digital or analog multimeter. In some cases the multimeter has a self-adjusting scale and no switch is necessary for this task.

Resistance measurements

The resistance measurement and the continuity test are the most common in normal work with electronics. To run a resistance test, apply a voltage to the circuit or device to be measured. The magnitude of this current causes the multimeter to display the value of its resistance.

Analog multimeters use an internal cell (normally a 1.5V AA cell) and digital multimeters use a 9V power supply. As the batteries in a multimeter lose power, so does the multimeter. Before each use, analog types must be adjusted to zero. This adjustment is shown in Figure 203.

Section 3—Troubleshooting and Repair

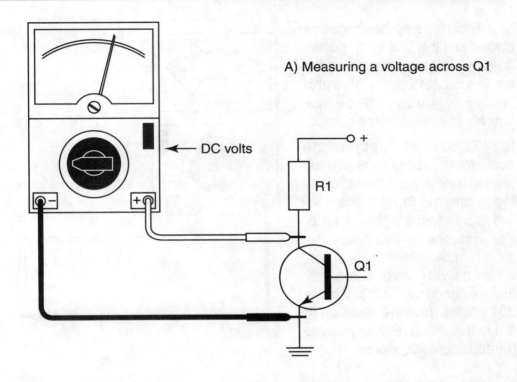

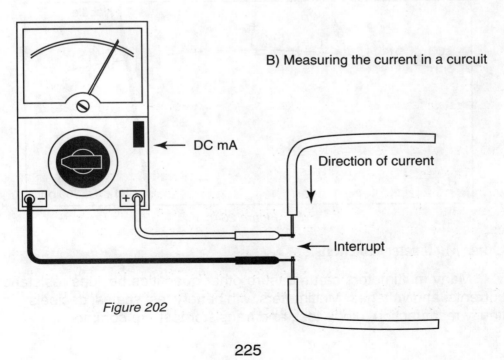

Figure 202

Place the probes together and adjust the "zero adj" potentiometer until the needle in the scale indicates zero. After this procedure you can use the multimeter to measure resistance.

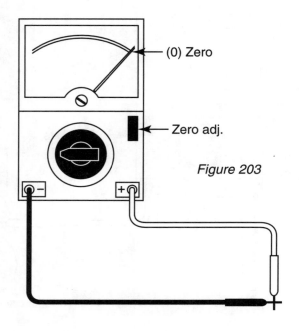

Figure 203

Choose the appropriate scale for the resistance you expect to measure. A better reading is one made from the central region to the right extreme. For instance, if you place the switch in the ohms x 1k scale, it means you have to multiply the readings by 1000. Figure 204 shows that an indication of 4.3 in this scale means you are measuring 4300 ohms.

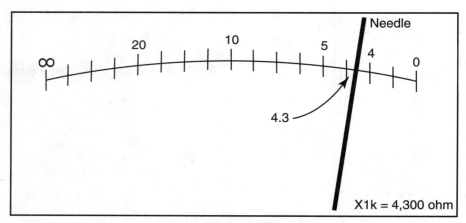

Figure 204

Other Multimeter Functions

Many multimeters can measure other quantities besides resistance, currents, and voltages. Multimeters with battery test scales, decibels, continuity test function, diode test, and transistor test are common.

Section 3—Troubleshooting and Repair

Reading and Interpreting Multimeter Values

Knowing that placing the probes of a multimeter in the terminals of a component and having it give you a reading either with a needle or LCD is not enough. You need to know what the reading means to conclude something of value. Some components are good when they present high resistance (capacitors, reverse-biased diodes, open switches, etc.) and some components are good with the reverse indication—when they present low resistance (fuses, wires, direct-biased diodes, coils, etc.).

At the end of the description of each component, the text explained how to test many components using a multimeter. When measuring voltages in a circuit, you must take care with interpretation of the values. Although the schematics indicate some voltages, depending on the multimeter, the presence of the multimeter in the circuit can change the values and give a false reading (mainly in less expensive analog types). Don't trust your multimeter! You must be sure that your multimeter can measure the indicated voltage in a circuit.

Finding Parts

Electronics today often are made up of one-way circuits. Once they present any problem, the entire board—including other components and parts—is replaced. Few components in equipment such as this can be replaced when problems occur. In some cases, the cost of replacing the components is actually higher than the cost to replace the board. The price of a technician's hourly work also figures in here. But, repairing old equipment or equipment not designed to be discarded when something inside is burned out, is possible and a technician must know how to do it.

After pinpointing the location of the cause and the components involved in the failure, the next problem to solve is finding a replacement. Today technicians can count on many component dealers, many of whom work through the Internet and mail order. Once contacted, most dealers will send the component(s) directly to your home, often in a few days, and in some cases, a few hours!

Electronics for the Electrician

For those with access to the Internet, some important dealer addresses in the United States and beyond are provided. Visit their web sites for more information about locations, shipping, and prices. Phone and fax numbers of many are also listed.

Dalbani
http://www.dalbani.com
U.S. Voice: 800-325-2264
U.S. Fax: 305-594-6588
Int. Voice: 305-716-0947
Int. Fax: 305-716-9719

MCM Electronics
http://www.mcmelectronics.com
U.S. Voice: 800-543-4330
U.S. Fax: 800-765-6960

Radio Shack
http://www.radioshack.com
U.S. Voice: 800-843-7422
U.S. Fax: 800-813-0087
Mail: Tech America, PO Box 1981, Forth Worth, Texas 76101-1981

Parts Express
http://www.parts-express.com
U.S. Voice: 800-338-0531
U.S. Fax: 513-743-1677

All Electronics
http://www.allelectronics.com
U.S. Order: 888-826-5432
U.S. Fax: 818-781-2653
U.S. Voice: 818-904-0524

Alfa Electronics
http://www.alfaelectronics.com
U.S. Voice: 800-526-2532
U.S. Fax: 609-897-0206

Other Sources for Components
- Arrow Electronics
- Avnet
- CalSwitch
- Digi-Key
- Electro Sonic
- Gerber
- Avnet Electronics
- JACO
- Marshall
- Newark
- Nu Horizons
- Reptron
- NetBuy
- Partminer SDirect
- Powell Electronics

Another important resource for components is old equipment that no longer functions. This equipment often has many parts that can be used as replacements. If you like to repair things and are always trying to recover old equipment, it can be interesting to have old boards or even seemingly ancient tube radios and amplifiers as a source of electronic components.

Transistors, resistors, and some types of capacitors are hard to destroy and can be found in good condition even many years after the original equipment is abandoned. The more critical components are electrolytic capacitors that can become dry after long periods of time, losing capacitance or going to a low-resistance state (short or loss) and can't be used. It is recommended to test all components before using them.

Equivalents

A common problem facing an electrician working with damaged electronic equipment is finding replacement parts when the original is not available anymore. When this occurs, an equivalent part is needed. However, these parts are not always exactly the same as manufacturer originals, and the same one cannot necessarily replace the same device in other equipment.

The most problematic cases involve semiconductors. Diodes, bipolar transistors, FETs, power-FETs, SCRs, triacs, and ICs come in millions of different types. Many manufacturers create their own types or designations to replace those made by competitors. In some cases they even add internal alterations, changing the basic characteristics for better performance. For example, an operational amplifier, such as the popular 741, can be found with designations LM741, 1741, uA741, LH101, CA3056, MC1539, and many others. Theoretically, they are the same component, but small internal differences can alter the performance and make one type improper for replacement depending on the application.

Transistors suffer from this problem as well. Many manufacturers used to adopt their own code when placing a transistor in equipment. The common 2N3904 or the BC548 can be remarked as, for example, 3X456-V-334-bis or something like this by a manufacturer. This makes the work of a technician replacing this component difficult. In many cases the intention is to force the user to look for an authorized service center in the event of problems.

When the component is common, it can be easy to find a replacement beginning with its characteristics. Experienced technicians, by simple observation of a circuit, often "know" which transistor is the ideal replacement type. As a general rule, the replacement type must have characteristics equal or superior to the original. If you are just starting out with electronics, it is not a good idea to attempt replacing pieces with these kinds of problematic designations.

Practical Circuits: How They Work

As pointed out in the beginning of the book, electricians today are finding electric parts not only in domestic, commercial, or car installations, but many electronic circuits. Another problem concerning an electrician relates to the electronic devices and appliances that operate from electric power and are sometimes in building installations. Although a deep knowledge about how these things work is not necessary to repair or install circuits, an electrician should have basic information about the operation. This information can be very useful when installing or troubleshooting an electric installation with a problem.

Let's look at the operation of some electronic circuits and devices found in electric installations and AC powered in homes, commercial buildings, and even cars. How do they operate and how are they placed in electric installations?

AC/DC Adapters

Many small electric and electronic appliances operate from DC power supplied from cells and batteries. In some cases, it is more convenient to run these devices from AC power to preserve battery life. AC/DC adapters or DC power supplies convert the high AC voltage present in the AC power line to a low DC voltage (in the range between 1.5 and 12V).

AC/DC adapters are formed by a simple circuit with a transformer to isolate the AC power line, avoiding shock hazards and stepping down the voltage to the desired value. Current follows the transformer leads to a rectifier (diode) and a filter (capacitor) in a typical configuration like the one shown in Figure 205.

Section 3—Troubleshooting and Repair

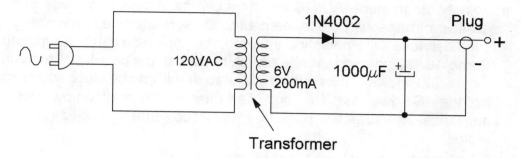

Figure 205

The output to the powered device is normally made by a cable with a plug as shown in the same figure. There are some details an electrician should pay attention to when using an AC/DC converter.

- Output voltage—output voltage can vary from 1.5V (one cell) to 12V (eight cells or a battery) for common AC/DC adapters. In some types, this voltage is fixed; you choose the voltage when buying the unit based on the voltage of the appliance. It is easy to identify this voltage from the number of cells. Each cell represents 1.5V, so if the powered device uses four cells it needs an AC/DC converter for 6V (4 x 1.5V). Some converters have a switch to choose the voltage. This switch can be adjusted based on the voltage of the powered device. It is important to make the correct adjustment because if it is a higher voltage than supported by the device, that device can be damaged.

- Output current—nominal output current is determined by the consumption of the appliance you intend to power. Normally, the consumption is evaluated by the size and type of the cells or batteries used to power it. In many of the cases, the current is indicated in the device, but if this is information is not available, use the next table to choose an AC/DC adapter.

Type of Cell	Current
AA	100 to 200mA
C	300 to 500mA
D	1 to 1.5A
9V cells	50 to 200mA

- Polarity of the plug—AC/DC adapters can be supplied by plugs with a positive or negative pole in the center. Be sure that the correct plug is used because if the polarity is inverted, the appliance can be damaged. Some AC/DC adapters have a switch to choose the polarity of the plug. Figure 206 shows a universal power adapter that can be plugged directly into the AC power line. This kind of adapter can be used to power small appliances, such as CD players, toys, video games, calculators, and radios.

An important point to observe is that common AC/DC adapters, in general, are very simple circuits with no voltage regulation stages. This means that the circuit presents a higher voltage in the output until the moment the appliance is plugged in, then it drops. This fact is important when testing an AC/DC adapter; the measured voltage in the output is normally higher than the adjusted or expected load until the device is powered on. When powering appliances where voltage is critical from the AC power line, it is important to use the specified types of adapters recommended by the manufacturer.

Figure 206
A Universal Power Adapter

Dimmers/Power Controls

Dimmers and power controls are devices used to control AC loads from an AC power line. They can be used to control the brightness of incandescent lamps, the temperature of heaters and small ovens, or the speed of motors including fans and pumps. The most common dimmer circuit is the one described in the Triac section starting on Page 162, where a sample configuration was analyzed.

This sort of circuit is usually small enough to be placed in the wall as a common switch, as shown in Figure 207. This dimmer can directly replace a wall switch. Equivalent dimmers using slide controls are also common and easily found for purchase.

Section 3—Troubleshooting and Repair

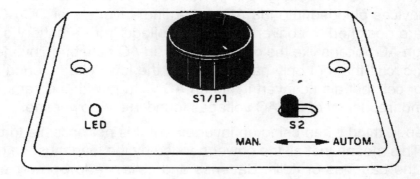

Figure 207

Another type of dimmer that is becoming popular is the touch dimmer. No mechanical parts are used to control the power applied to the load in this dimmer. The touch controlled dimmer uses a small metal plate where, by the touch of your fingers, the power applied to the load changes, slowly rising or falling to the desired level.

There are some special integrated circuits that have been designed for the task of operating a device with touch. One of them is the SLB0587 from Infineon, which contains all the processor circuits to slow up and slow down the power applied to the load from the touch in a sensor.

Dimmers can replace any wall switch in basic applications. When working with a dimmer an electrician must be attentive to some details.

A. Be sure the dimmer can control the load. Fluorescent lamps can't be controlled by common dimmers. Electronic appliances, such as audio amplifiers, TVs, VCRs, and some others can't be controlled by a dimmer.

B. Invert the positions of the wires in a dimmer if it doesn't control the load properly.

C. Use filters (see the sections on SCRs, triacs, and EMI for more information) if the circuit causes interference in other electronic equipment.

Electronics for the Electrician

Inverters

Batteries are low-voltage DC power supplies and can't be used to power devices plugged into the AC power line. Inverters or DC/AC converters are designed to convert a low DC voltage from a battery or cells into a high AC voltage like the ones found in an AC outlet. An inverter is a complex circuit that not only has to step up the low voltage found in the battery or cell, but also convert it into an AC voltage with the same wave shape and frequency of the AC voltages found in the power line.

Keep in mind when using an inverter that it is not possible to create energy. This means that all the power needed by the load must be generated by the batteries or cells. Batteries and cells are low-power energy sources, which means that the amount of power sourced by an inverter can't be high, or under any circumstances, higher than the capabilities of the cells or battery. As a general rule, inverters powered from cells or batteries are only used to power low-consumption appliances, such as large fluorescent lamps in emergency systems, signaling devices, and xenon beacons. Inverters powered from car batteries or other heavy-duty batteries can source energy to medium-power appliances like small TVs, electric razors, small fluorescent lamps, transmitters, laptop computers, small xenon beacons, and incandescent lamps.

How they work

A typical inverter is formed by a low-frequency oscillator running on the AC power-line frequency and a high-power output stage using transistors (bipolar, power FETs, or IGBts) to drive a transformer, as shown in Figure 208.

The signal applied to the low-voltage winding of the transformer appears as an AC high voltage in the secondary winding. The quality of the transformer determines how much power can be transferred from the low-voltage to the high-voltage circuit, determining the efficiency of the converter. Values as high as 90 percent are common in commercial types, meaning that for each 10 watts of power drained from the battery, 9 watts are load and 1 watt is transformed into heat by the circuits.

Most of the heat is converted by the output transistor because they are mounted in large heat sinks. In commercial types, the metallic box is used as a heat sink.

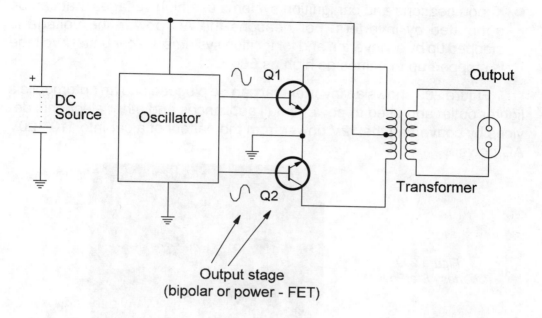

Figure 208

It is important to observe that in many types, the wave shape of the output voltage is not sinusoidal. The conversion made by circuits using pulses and the construction of the transformer can result in different wave shapes. In the applications here, the wave shape is critical because it is important to see if the inverter is working as needed. For an electrician, the inverters are important because they are found in emergency lighting systems or powering devices that are normally powered from the AC power line.

The following are some important applications for inverters.

- They can be used to power electric appliances from batteries and cells in emergency situations or in places where AC power is not available, such as cars, boats, and on camping trips.
- Inverters for fluorescent lamps are used in emergency lighting systems. The batteries used in these systems are charged by AC power-line voltages.
- They are used in no-break systems for computers, which supply energy for a computer if an AC power outage occurs.

Electronics for the Electrician

- Xenon beacons and car ignition systems use high voltages that can be generated by inverters. For beacons the AC power-line voltage is stepped up by an inverter, and for ignition systems the car battery voltage is stepped up to values as high as 600V.

Figure 209 shows an inverter that can be plugged into an automobile's lighter outlet and used to power small appliances and other electronic devices by converting the 12V power from the battery of a car into 110-120V AC power.

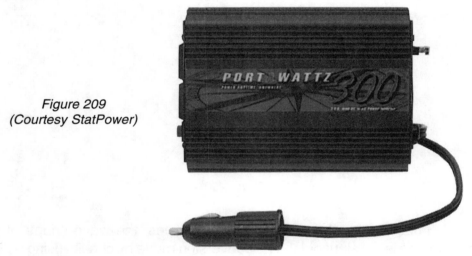

Figure 209
(Courtesy StatPower)

Working with inverters

It is very important for an electrician to know what kind of inverter is recommended for an application. Not only are inverters critical equipment, but also the powered circuits can be damaged if an incorrect choice is made. The following are main specifications an electrician must observe when working with inverters.

A. Output power

It is important to be sure that the inverter can supply enough power to the load. The power specification is given in watts (W). Take care with audio equipment where the output power doesn't reflect the consumed power; it can be given in a unit of watts, peak, or PMPO (instantaneous power peak).

B. Wave shape

The wave shape of many inverters isn't sinusoidal. This fact can be important in some applications. Lamps and heaters don't need a sinusoidal wave shape to be powered.

C. Performance

The higher the amount of energy converted, the better the inverter. Good inverters have a performance of 70 percent or more.

D. Isolation

The high voltages found in the output of inverters can cause severe shocks. Appropriate wires must be used to connect the loads if they are not plugged directly into an outlet, as in the case of fluorescent lamp inverters.

E. Battery placement

If not sealed, batteries used to power the inverter must be placed where the poisonous gases produced can be eliminated.

F. Wiring

Wiring to the battery can be done using wires based on the current drained by the inverter. A 100W inverter with an efficiency of 80 percent can drain a current as high as 12A from the battery. The wires must be able to handle this current.

Troubleshooting

The output transistor is the most fragile component in this type of circuit because it operates under limited conditions. When replacing these transistors, make sure that the type chosen supports the same voltage. The suffix indicates, in many cases, the operational voltage. For example, a TIP32A supports less voltage across it than the TIP32B.

Flashers

Lamp flashers are used in signaling. Garage doors, signal lights for customers in stores, and warning devices are some cases where flashers

Electronics for the Electrician

can be used. The simplest electrical type uses an incandescent lamp with some thermo-mechanics to open and close the circuit. But, with the aid of electronics circuits, the electrician can find many different configurations for lamp flashers.

A. Incandescent

Low-voltage flashers, such as the ones used in cars or powered from DC supplies, can be made with simple configurations like the one shown in Figure 210.

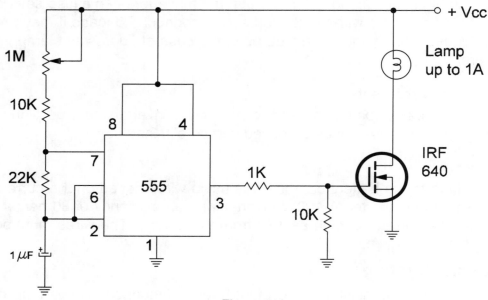

Figure 210

The power transistor (bipolar or FET) controls a lamp that flashes in a rate determined by the circuit adjustment. To flash a high-voltage, AC-powered incandescent lamp, a circuit using an SCR or a triac is more suitable. Figure 211 shows a circuit that can be found in garage doors using two lamps. The lamps flash in an alternating mode and the frequency is determined by an internal adjustment of the circuit.

In both cases the most sensitive of the components are the transistors and the SCRs (or triacs) that can open or short. When replacing these components, an electrician must observe the operating voltages.

Section 3—Troubleshooting and Repair

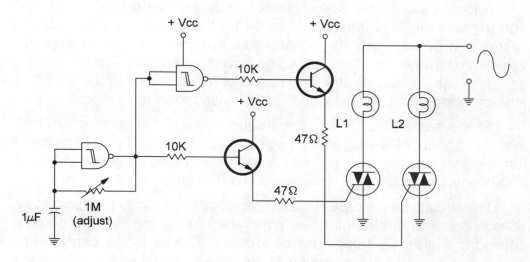

Figure 211

B. Xenon (strobe)

The xenon beacon or xenon flasher is used on towers and high buildings as a signaling device. Portable xenon beacons can also be used in emergency situations (powered from cells or batteries) or in police and rescue vehicles. The basic circuit of a battery-powered xenon beacon is shown in Figure 212.

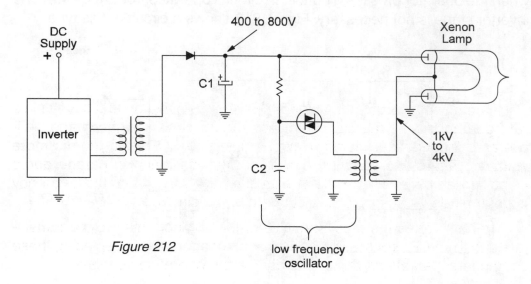

Figure 212

Electronics for the Electrician

A low-frequency oscillator operates as an inverter driving a high-voltage transformer. This transformer steps up the voltage to the necessary values in order to power the xenon lamp. For common xenon lamps the values are between 400 and 800V. The high voltage is rectified and used to charge a high-value capacitor. This capacitor (in some circuits many capacitors are wired in parallel forming a bank) is used to store electric energy.

At the same time, a very low-frequency oscillator produces pulses that are used to drive another high-voltage transformer. This second transformer has a secondary winding rated to voltages between 800 and 2000V—values necessary to trigger the xenon lamp.

The circuits can use SCRs, transistors, or even ICs to produce a low-voltage pulse to the transformer. When the high-voltage trigger pulse is applied to the lamp, it becomes conductive. The capacitor can then be discharged across it, transforming all the stored energy into a high-power, short light pulse. As long as the capacitor is discharged, the lamp becomes nonconductive again and a new operation cycle begins.

The output power, or luminous power, of a xenon beacon is measured in Joules (J), which is the energy stored by the capacitors. Typical values are in the range of millijoules (mJ).

The operation principle of these xenon flashers is the same as camera flashes. The difference is that, in the camera flash the circuit is triggered when the operator presses a button. In circuits operated from AC power, the inverter stage is not necessary. Figure 213 shows a circuit of this type.

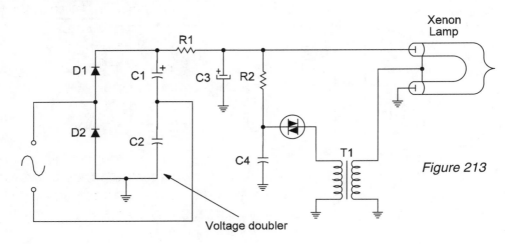

Figure 213

Section 3—Troubleshooting and Repair

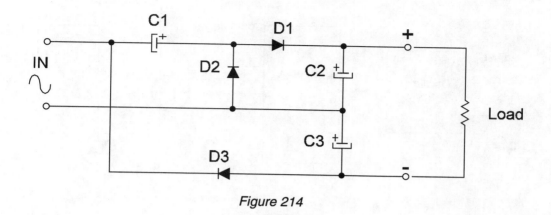

Figure 214

A voltage multiplier or a transformer is used to step up the AC voltage that can raise the necessary values for the lamp operation. A voltage multiplier is a network formed by diodes and capacitors that can step up the AC voltage converting it to a higher DC voltage as shown in Figure 214.

The important parts in this circuit are the diodes and capacitors that store energy. When working with these circuits an electrician must observe some specifications.

- When replacing the lamp, be sure that the replacement type has the same triggering voltage and the same power (mJ).
- Xenon flashers run with AC power are not isolated and can cause severe shocks if live parts are touched.
- In all devices, after the inverter stages, very high voltages can be found. These voltages are present even after the circuit is no longer plugged into the AC power source, as the voltage is stored in the capacitors. Do not touch the capacitors even when the device is unplugged.

C. Sequencers

Another type of luminous signaling device is the sequencer. This kind of device is found in the form of arrows indicating emergency doors or emergency exits and is even used in some types of Christmas tree lights. Figure 215 shows a typical sequencer circuit for incandescent lamps (or LEDs) that can be used in some signaling appliances.

Electronics for the Electrician

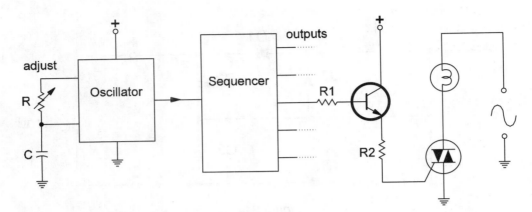

Figure 215

The clock circuit is an oscillator that determines the speed of the effect. The signals from this block drive the counter or another circuit that decodes the desired sequence or effect, such as a microprocessor or microcontroller. The decoded outputs of the sequencer circuit drive a transistor if low-voltage DC loads are controlled, such as LEDs or small incandescent lamps. If high-voltage AC loads are driven, SCRs or triacs are used in typical configurations like the ones shown in Figure 216.

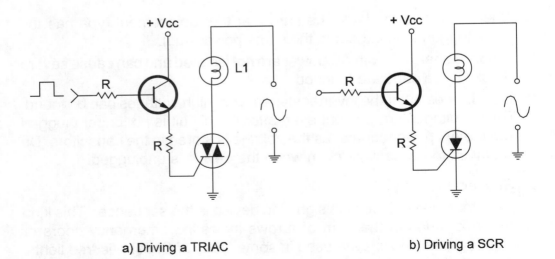

a) Driving a TRIAC   b) Driving a SCR

Figure 216

Section 3—Troubleshooting and Repair

When using this type of circuit the electrician must keep in mind that the control block is a low-voltage circuit. The power circuit depends on the type of lamps used in the effect.

Automatic Lighting/Emergency Lighting

Some years ago, the common way to control a lamp—turning it on and off—was using a switch. With electronics today, many automatic ways to control lamps have appeared in electric installations. One of these controls is the automatic lighting system. This system is very useful in gardens, house entrances, and other areas where timed lighting is desirable. An automatic light turns on a lamp at dusk and turns it off at dawn. This kind of control is found in many formats from lamps plugged directly into the socket of a common incandescent lamp (Figure 217) to more complex systems that can control all the lamps in a garden or on a street.

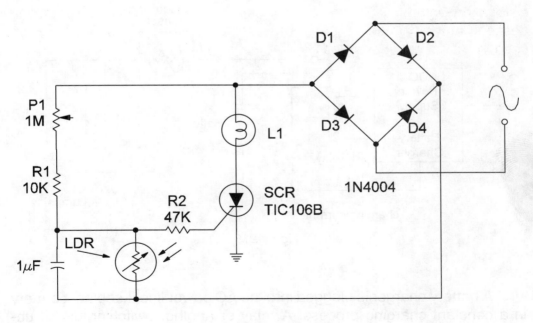

Figure 217

Electronics for the Electrician

The sensor is an LDR or cadmium sulfide photocell that is mounted in such a way so that it receives only the ambient light and not the light from the lamp it controls, thus avoiding feedback. When the amount of light received by the sensor falls, the circuit triggers an SCR or triac that turns on the lamp. As with any circuit using SCRs or triacs, this kind of control can generate some radio interference (refer to the section on SCRs, triacs, and EMI).

Depending on the application, the electrician must install filters in the AC power line into which the device is plugged. When working with these devices, it is important to be sure that the sensor only receives ambient light. If light from the lamp is picked up by the sensor, a feedback process can make the circuit unstable.

Emergency lights turn on one or more lamps when the AC power fails. These lights are required equipment in public places, such as restaurants, movie theaters, schools, and many others. The basic emergency lighting circuit has an operation principle as shown in Figure 218.

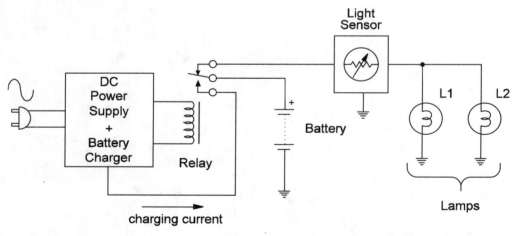

Figure 218

A battery charger is plugged into the AC power line, keeping a battery in a constant charging process. A relay or another switching circuit unplugs this block from the battery when the AC power fails and, at the same time, plugs the battery into the emergency lamps. A photocell is placed

Section 3—Troubleshooting and Repair

between the blocks to give the circuit some "intelligence." This block only turns on the lamps if the location is dark. It is not necessary to turn on the lamps during an emergency if the area is illuminated by natural light. In some systems there is an inverter between the battery and the lamps. This inverter is necessary to drive fluorescent lamps.

Alarms

Home security is increasingly becoming more important. High-tech equipment is available and can easily be installed in any home. The degree of sophistication of systems can vary from simple switches that trigger an alarm if a door or window is tampered with to complex microprocessed systems that dial a security provider's or the police's phone number if some disturbance is detected. Electricians are sometimes called to prepare a home to receive some kind of alarm system, which may include placement of sensors and wiring. They must also install the power supply for the system, which includes outlets, wires for sirens, and other alarm transducers. So, it is important they know how the basic system works to help in the installation or troubleshooting. The following are some of the basic alarm and security systems.

■ Magnetic sensors and switches

The simplest alarm system uses a switch to trigger a siren or another sound or light source. Figure 219 shows a diagram of a system that uses a magnetic switch placed in a window.

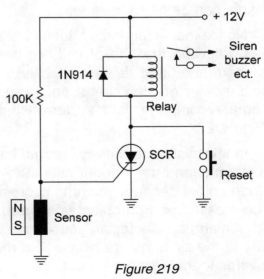

Figure 219

With the window closed, the magnet is in contact with the reed switch, keeping the contacts closed and the circuit off. If the magnet is moved away from the reed switch, the contacts open and the

Electronics for the Electrician

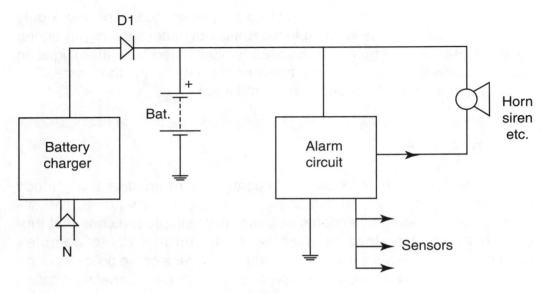

Figure 221

alarm is triggered. Many switches of this kind can be wired in series to protect all the windows and doors of a house or building. The relay is open until one of the sensors is activated. Once triggered, even if the window or door is closed again, the circuit remains on. To stop the alarm it is necessary to turn off the power supply, pressing S1 for a moment. Any number of sensors can be wired in series.

Notice that in this circuit, the current flowing across the sensor is very weak, allowing the use of cells or batteries for power. Radio Shack , Dimango, and other dealers offer the sensors for this type of alarm, and also the detectors, the sirens, and in some cases, complete kits to protect a home. A complete alarm system includes additional resources as shown in Figure 221.

In this system, a battery charger keeps the battery charged to power the alarm even if an intruder cuts the AC energy supply. The circuit also includes other sensors, such as microswitches (small switches activated when pressed or if an object is moved). When installing or repairing this kind of alarm an electrician doesn't need to be extremely experienced as many of the systems are simple, but must take care with the isolation of the wires to the sensors.

Section 3—Troubleshooting and Repair

Another system includes in the sensor a radio transmitter that sends a signal to a receiver. If the sensor is activated, the signal changes and the alarm is triggered. This system is suitable for cases where there is no space to pass common wiring.

■ Photoelectric

The use of photo sensors in alarms is common in many applications. Figure 222 shows a passage alarm or passage detector using a phototransistor as the sensor and an IR (infrared) LED as the radiation source.

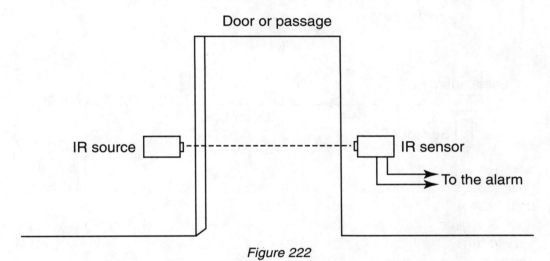

Figure 222

This circuit detects when a person or object cuts the radiation beam, even for a fraction of a second, in the sensor. The use of IR is recommended as the intruder can't see it, and therefore can't see where the system is placed.

Many circuits operate with modulated IR, meaning the circuit doesn't produce a continuous light beam, but produces short pulses in a certain frequency to be detected by the sensor. This procedure is used to avoid an intruder using a "false" IR source to stall the alarm as suggested in Figure 223.

When installing or working with this kind of device, an electrician must be careful with light sources that can cause interference in the system, such as fluorescent lamps. If large areas must be protected, the alarm can use a laser as its IR source (Figure 224).

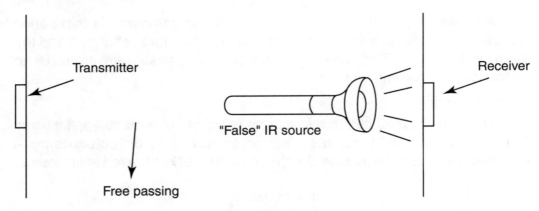

Figure 223

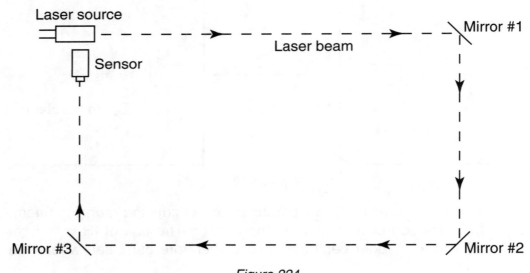

Figure 224

Mirrors can bend the IR laser beam, fencing a closed area against any intruder. If the laser beam is disrupted on any side of the area, the alarm will be triggered.

■ Motion Detectors

Motion detector alarms use a piroelectric sensor. It is formed by piezoelectric material that presents a surface electric charge that changes according to the amount of IR radiation falling onto it. Because a certain degree of IR radiation always exists, the circuits are designed to detect

Section 3—Troubleshooting and Repair

small changes in this radiation from the presence of a warm-blooded creature, such as a human. To detect the very small changes in the radiation, the circuit used is extremely sensitive and can be complex, involving special ICs.

- Vibration Alarms

Small vibration sensors are placed in home or shop windows to detect any type of impact or vibration. The circuit for this type of sensor is the same as in the magnetic switch.

- Ultrasound

In the ultrasound or super-sound system, the air is filled with inaudible sound vibrations (above 20,000 Hz). A sensor is placed to monitor these vibrations. The system makes an "image" from the pattern of direct and reflected sound waves and any alteration due to movement of an object—an intruder—when detected by the circuit. The sensors and the emitter must be placed in the positions indicated by the manufacturer as it is sensitive to external interference.

Doorbells And Chimes

The traditional electric doorbell is formed by a coil (electromagnet) and a metal plate that vibrates to produce sounds when powered by AC voltage. Modern doorbells and chimes use electronic circuits to produce musical sounds, such as notes in sequence, melodies, and even bird songs. In some cases, sequential systems act on solenoids that beat on metal tubes with varying lengths.

The basic circuit of an electronic doorbell in blocks is shown in Figure 225.

When triggered by the doorbell button, the circuit runs a program that produces a sequence of notes programmed in the chip. The signal corresponds to the notes and is amplified by a common low-power audio amplifier. Audio amplifiers with output powers in the range between 1 and 5 watts are common in commercial circuits.

The way this kind of circuit is installed is not much different from any common doorbell. The trigger button (pushbutton) is placed near the door.

Electronics for the Electrician

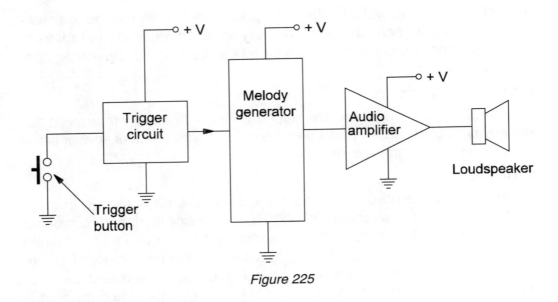

Figure 225

Many types allow more than one trigger point. The circuit is powered from the AC power line with a transformer to supply DC power. The power supply providing the low voltage to the DC circuit needs to remain on all the time. Some types of doorbells and chimes are coordinated with intercoms and video cameras.

Battery Chargers

Two types of battery chargers are used depending on the batteries to be charged. One is the small type used to charge Ni-Cad batteries and cellular telephone batteries or those embedded in wireless telephones and intercoms. This circuit is nothing more than a low-voltage DC power supply with a circuit that supplies constant current to the cells to be charged. The basic idea is that, by forcing a current to flow in the inverse direction of the normal current supplied to the load, a battery is recharged (Figure 226).

In this circuit, the resistors determine the amount of current used to charge a battery based on its type. This means that battery chargers have specifications indicating what type of battery they can accommodate. If the

Section 3—Troubleshooting and Repair

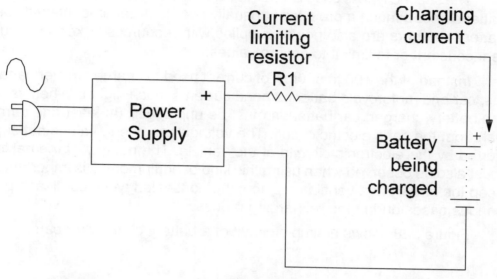

Figure 226

incorrect current is used, the battery life can be reduced. The most common types are the ones used to charge AA, C, D and 9V, and ni-cad batteries (Figure 227).

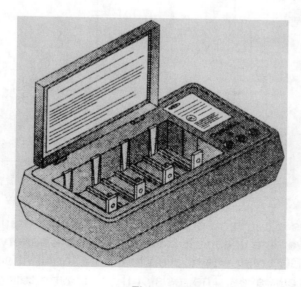

Figure 227

The second type is used to charge car and motorcycle batteries. These batteries need much more current usually over a larger time interval. Car battery chargers are also power supplies with resources to keep the current constant or to limit it to secure values.

Instead of the 100 to 500mA of current used to charge smaller batteries, car and motorcycle batteries need currents much higher. When using the battery charger, car batteries must be unplugged (at least one terminal) from the car's electric circuit. The voltage can rise to values not supported by the electronic circuits of the vehicle. The polarity of the cables must also be observed when using this kind of equipment. It is important to keep the battery in a ventilated place; the nonsealed types can leach poisonous gases during the recharging process.

Figure 228 shows a simple circuit of a battery charger for cars.

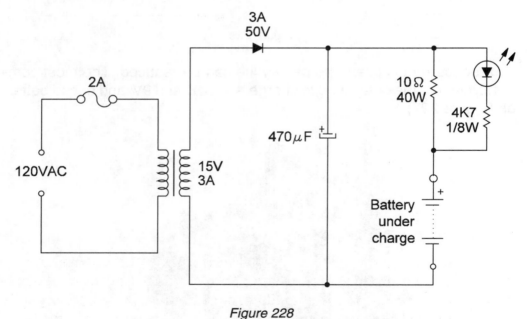

Figure 228

This circuit is a common power supply where the transformer steps down the AC power-line voltage to a value above the battery voltage. After rectification, the DC voltage is applied to the battery in the inverse direction of its normal operation. The resistor (R) limits the current to a secure value for the charge, and the LED monitors the charging process.

Section 3—Troubleshooting and Repair

The amount of time to charge a common battery can vary from two to three hours to start an engine, and more than 12 hours for a complete charging process. If the battery doesn't retain the charge, it must be replaced. Sophisticated circuits can include resources to detect when the battery is completely charged and automatically interrupt the charging process.

Fluorescent Lamps

In traditional installations, fluorescent lamps use a starter and a ballast in a configuration, such as shown in Figure 229.

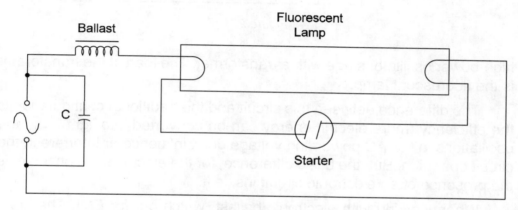

Figure 229

The starter acts as an automatic switch closing and opening the inductive circuit to produce high-voltage spikes. These spikes are necessary to ionize the gas inside the fluorescent lamp. As long as the gas is ionized, the lamp becomes conductive, and light is emitted. When the lamp is conductive, after the initial process, the starter is now inactive.

With electronics, the traditional ballast is replaced by high-voltage and high-frequency circuits that use modern components like power FETs and ICs (Figure 231). This circuit is really a high-frequency inverter operating directly from the AC power line. The AC is rectified and used to drive a

Electronics for the Electrician

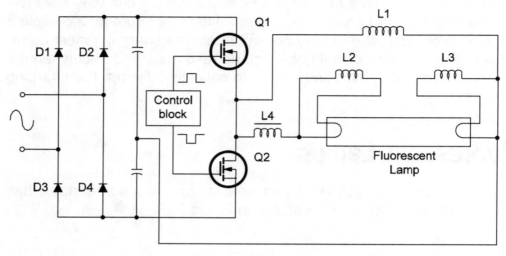

Figure 231

high-power oscillator stage with a transformer. The load of the transformer is the fluorescent lamp.

The difference between this circuit and the traditional configuration is the efficiency (more electric energy can be converted into light), and the oscillations in the AC power-line voltage don't influence or interfere in the circuit operation. But, the basic difference, for the electrician to observe, is the presence of an electronic circuit inside it.

When working with electronic ballast, watch out for EMI. The high-frequency operation of some circuits requires special care to avoid interference with nearby radio equipment. The basic procedures include the use of filters and grounding.

AC/AC Converters

The difference in AC voltages in many countries can be a problem when you have an appliance not able to operate on more than one specified voltage. If you purchased an appliance in another country, or if you are moving to another country, you must consider that your appliances may not be able to plug into an outlet.

The use of a step-up or step-down converter that can convert the 120VAC power-line voltage to 220 or 240VAC or a 220/240VAC to 120VAC is recommended. If you plug a 220/240VAC appliance into a 120VAC outlet, it won't work; however, if you plug a 120VAC appliance into a 220/240VAC outlet, the appliance may burn, have parts damaged , or have a fuse blown.

When recommending a step-up or step-down transformer, electricians must observe the power needs of the appliance. The power is measured in watts (W), meaning you have to use a transformer with more power than required by the load, with a certain tolerance margin to avoid overload. Many step-down transformers used to convert 220/240VAC into120VAC can also be used to convert 120VAC into 220/240VAC when reverse wired to the appliance. This is not a general rule because it is valid only for simple transformers without voltage regulation circuits included inside.

Power Conditioners

In theory, the AC power-line voltage must be kept constant with the 120VAC or another value specified by the energy company or according to the country. Unfortunately, local conditions, as consumption peak hours occur or long wires are used to power an appliance, the voltage can step down to unacceptable values.

Many modern appliances have circuits that include voltage regulators that let them operate in a large range of voltages without problems. But, if this type of resource is not found in the appliance, an external voltage regulator transformer or power conditioner is necessary. An external voltage regulator is an electric/electronic device connected between the appliance and the AC power-line outlet with the purpose of maintaining constant voltage. The most common types are automatic, using electronic circuits that sense the output voltage and correct it by acting on a transformer. When using this kind of device, it is important to check that the specified power is enough to provide the load with the necessary energy.

Intercoms

Many types of intercoms can be found in the electric installations of homes or buildings. The main type is the one associated with the doorbell and used to talk with a person at the front door. It is formed by a low-power audio amplifier where the loudspeaker is used to both reproduce the sounds and to act as a microphone. Figure 232 shows a simple circuit of an intercom using an IC from National Semiconductor.

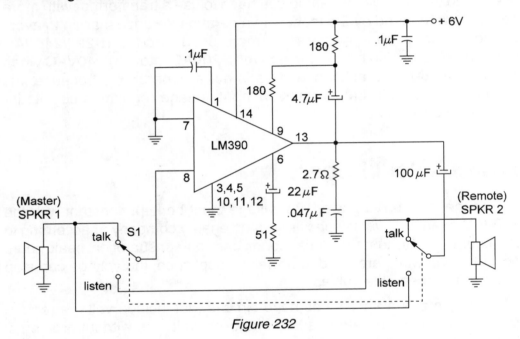

Figure 232

The switch changes the function of the loudspeakers, plugging them into the input when they are used as microphones (to talk) and as loudspeakers (to hear). Other types can use separated elements for these functions, and electret microphones are used.

The same amplifier can also be used to reproduce the sound of the doorbell. In the basic installation, the central unit is placed inside the building and powered from the AC power line, cells, or batteries. The remote unit is placed at the front door or entrance of the building and two types of wires can be used. In some cases, separated wires for the bell and audio

Section 3—Troubleshooting and Repair

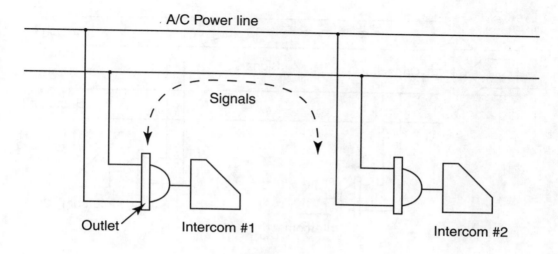

Figure 233

signals are used. The audio signals, in some cases, are conducted by screened cables.

Another type of intercom is one that uses the AC power line to transport the signal between the stations. This means that you only need to plug the units into outlets (Figure 233).

High-frequency signals are superposed to the voltage transporting the audio signals. Common types use PLLs to decode the signals in the range between 50 and 200kHz (refer to the section on PLLs). The principal problem found when working with this kind of intercom is that in some cases the units are plugged into outlets of different branches in an electric installation, as shown in Figure 234.

Even in adjacent rooms, the signals must travel a large distance. Sometimes they are unable to pass from one branch to another. The presence of inductive elements between the branches of the installations can stop the signals. Installing a 0.1 uF/400V paper or polyester capacitor between the branches where the signals can't pass will solve the problem.

Many versions of commercial intercoms can be purchased in many specialized stores and easily installed by an electrician, even one just beginning.

Electronics for the Electrician

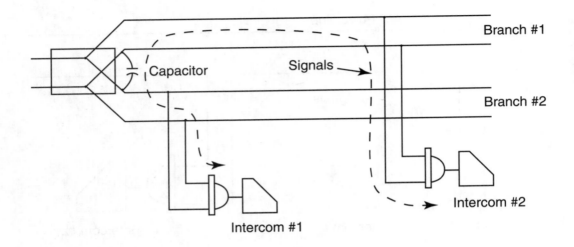

Figure 234

Wireless Systems

The use of the AC power-line voltage is the simplest solution for remote control of appliances and also the most traditional. By placing a switch at any point in the wiring, you can turn on and off any appliance plugged into that line. Today, with the use of complex devices—many of them as small as a matchbox—it is possible to control appliances without wires or the AC power line.

Many solutions have been found to make the control of electric and electronic appliances using electronic circuits easier. The electrician who installs appliances in any building that are AC powered should know how they work. This is especially important to avoid problems with correct installation, interference, or improper use. For example, a remote control for a garage door that uses radio waves in a location with a high level of interference doesn't work well. It can be replaced by an IR remote control with great success.

An electrician can find the following types of remote control for domestic appliances.

Section 3—Troubleshooting and Repair

■ Radio Frequency

This kind of remote control uses a high-frequency transmitter operating in the VHF or UHF band (between 30 and 600MHz) to send signals to a receiver placed at distances of up to 50 meters away, depending on the application. The signals are modulated, (i.e., transport information) communicating to the controlled device what it has to do. The simplest systems use a transistor as an oscillator to produce the high-frequency signals and an IC to produce the coded information to be sent. In some cases, the transistor is replaced by hybrid modules or a complete transmitter mounted in small modules. Figure 235 shows a typical transmitter used to open and close garage automatic doors.

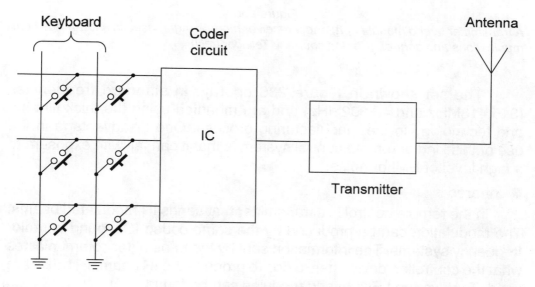

Figure 235

The IC can send a coded signal in a manner so that the receiver only recognizes the signals sent by that IC. The receiver decodes the signals, acting on relays, motors, and other devices to be activated. Normally, in the receiver we find the other member of the pair of ICs used to aid in decoding the signals. Small modules, or hybrid modules, forming pairs of transmitters and receivers are found in many remote control systems, such as the ones shown in the Figure 236.

Electronics for the Electrician

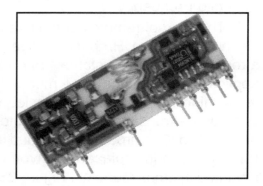

Figure 236
A transmitter hybrid module (left) and a receiver module (right) used in short-range communications and control. (Photos courtesy Tellecontrolli)

The set shown in Figure 236 operate in standard frequencies (315,418MHz and 43,392MHz) and are mounted using the thick film hybrid technology in the manufacturing process. One disadvantage in the use of this kind of remote control system is that it can fail with exposure to a high level of EMI or noise.

■ Infrared

In the remote control is a transmitter that sends IR modulated beams. The modulation can be produced by the same coded ICs found in radio-frequency systems. The information sent by the transmitter communicates what the controlled device has to do. To produce the IR beam, IR LEDs are used. To drive the LED, hybrid modules can be found.

In the receiver, a phototransistor or photodiode exists acting as a sensor for the IR beam. The signal produced by this receiver is applied to a decoder circuit (in some cases the same as found in radio systems) that decodes the information to the driver circuits. This sort of system is found in many electronic appliances, such as VCRs, TVs, and audio systems. It is also found in electric appliances, such as automatic garage doors, remote-controlled fans, and automatic doors.

Although the IR is not disrupted by interference by noise sources like RF or radio systems, its principal disadvantage is that the receiver must "see" the transmitter, meaning the units must have a direct line of communication open. Any solid obstacle can block the IR beam.

The range depends on local conditions, but normally doesn't exceed 50 meters.

- Ultrasound

Ultrasound vibrations can activate some appliances. This system is no longer in use, but it can be found in old appliances including TVs. The transmitter contained a small ultrasound transducer that sent vibrations in frequencies near 42kHz. These vibrations were picked up by a transducer tuned to the same frequency. The signal was also modulated by a circuit that sent information to the receiver, relaying what the appliance was to do. Due to its disadvantages, when compared with other modern systems, such as the IR and radio frequency systems, it is no longer used.

- AC power line

The AC power line can be used to send signals from a transmitter to a remote receiver connected to an appliance. The signals travel through building wiring, which transports information about what the appliance has to do. The operation principal is the same as found in wireless intercoms that use AC power to transport their signals. Although some appliances use this system, its performance is limited to short-range applications.

- Low-current wiring

Sometimes heavy-duty appliances, such as air conditioners, heaters, and high-power motors must be turned on and off remotely. If the control is part of the power cord, large-gauge wires will be needed because of the large current needs. Even the switches must be heavy-duty types, increasing the size of the control unit and adding a danger of shocks and shorts. A solution to this problem is the use of units having relays activated or controlled by low-current circuits like the one shown in Figure 237.

The current used to trigger the relay on and off is very low (milliamperes), allowing the use of low-gauge wires. And, as the control circuit is isolated from the AC power line and the voltage is very low, shock hazards are avoided.

On the Internet, an electrician can find many types of remote controls that can be adapted to common electric and electronic appliances. When

Electronics for the Electrician

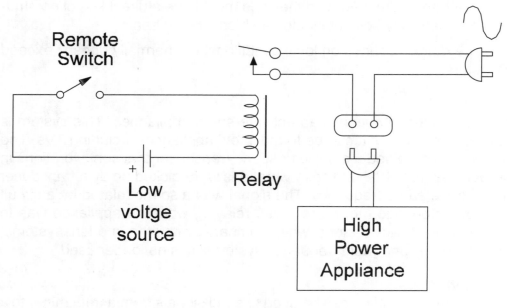

Figure 237

choosing one for an application, keep in mind if the area is noisy, with many interference sources. Avoid radio systems and use IR, or if there are obstacles choose a radio system over IR.

High-Frequency Appliances

Many electric and electronic appliances have high-frequency stages in their circuits. When installing or working with those appliances, it is important to know that radio frequency signals are critical and can be a source of problems for the less experienced electrician.

The first case to be considered is the installation of external antennas for TVs in the traditional VHF or UHF bands. Although the AC power-line voltage can be sent to every part using common wires, this is not a rule for high-frequency signals. High-frequency signals are attenuated by large lengths of wire and can be sensitive to bends or the close proximity of metallic objects.

A single curve in a wire can be sensed as a coil or inductance for a high-frequency signal. Two wires, one passing near the other, can represent a capacitor for high-frequency signals, causing alterations in its strength as shown in Figure 238.

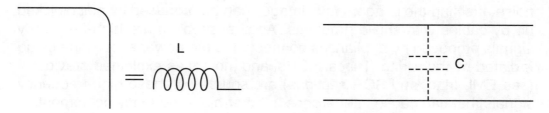

Figure 238

Another important point to consider is the fact that the wiring of a building acts as an antenna, radiating radio signals in the VLF (very low frequency band). These signals can be picked up by any nonshielded wires and, if amplified by audio circuits, can be reproduced as a hum in loudspeakers or phones. This problem can be noticed in audio amplifiers when you touch the input jack as shown in Figure 239.

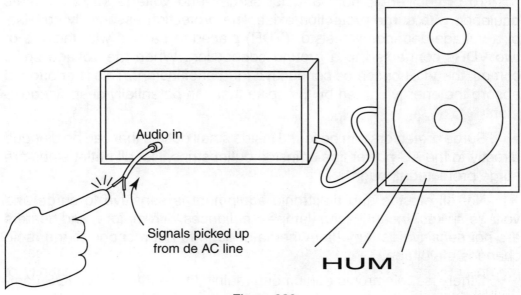

Figure 239

In audio systems and intercoms, the hum can become a serious problem for an electrician trying to install this kind of equipment. Shielded cables only work if the screen is wired to a good ground. So, it is important to plan an installation for these kinds of equipment.

In video systems, an incorrect connection for the cables can result in noise affecting the image. Wavy images can be produced by noise picked up by cables with shield problems. Another problem are high-frequency signals produced by appliances connected to the power supply line being radiated by the wiring. Triacs, SCRs, and motors as explained previously (see EMI, triac, and SCR sections) are sources of these high-frequency signals that can cause interference in other high-frequency equipment.

Surge Protectors

Brief pulses of voltages that can exceed the normal 120 or 220/240V (depending on the voltage used in the AC power line) are called surges. These pulses can rise, in some cases, to volts as high as 1000 and can destroy equipment. Sometimes these pulses are produced by nearby lightning strikes.

To protect equipment against surges and voltage spikes, outlets equipped with surge protection exist. The protection essentially consists of a voltage-dependent resistor (VDR) placed in parallel with the line or two VDRs connected to a ground connection. When the voltage spike comes, the VDR becomes conductive for a brief instant. This is enough to absorb the energy carried by the spike that can potentially destroy equipment.

Surge protectors can be found inside small outlets that can be plugged directly to the AC power line or inside outlet strips. Not all outlet strips are surge protected, though.

Not all electric and electronic equipment is sensitive to surge and voltage spikes. Incandescent lamps, appliances with motors, and heaters are not sensitive as they have inertia enough to not respond to the rapid changes of voltage.

Otherwise, electronic equipment, mainly those that use CMOS components, is very sensitive to voltage spikes coming through the AC power line. This includes computers, electronic telephone answering machines, VCRs, TVs, and audio equipment. Telephone equipment and computers connected to the Internet by telephone lines are also sensitive to voltage spikes coming through the telephone wiring.

Although they have internal surge protection circuits, in some cases they are not fast enough to handle the spikes, or the spikes are strong enough to not be absorbed by them. So, it is common to find modems and other such equipment destroyed after a severe storm.

It is a good idea to connect all this type of equipment to the AC power line using protected outlets or to unplug them from the telephone line or AC power line during storms with severe lighting.

Conclusion

This book is only a brief introduction to electronics for the electrician. There are so many more things to learn and know about electronics as it relates to electric installation. The buildings are becoming even more sophisticated with electronic equipment being embedded in the electric installations. The electrician of the future will have to know not only about the traditional AC power-line voltages, but should have a broad base of knowledge in many complicated areas including digital electronics, microcontrollers, and DSPs.

The idea of wiring a building to connect with the Internet, to send information to a remote computer, or to have climate conditioning systems plugged into sophisticated computers that can sense humidity and temperature, are not old. The electrician of the future, the construction engineer, and even the architect without some knowledge of electronics will be in an unfortunate situation. Even if you are not directly related to a profession that works with electricity, electronics are present in almost all things we make. To know something about it is important to our comfort, our performance in any job, and even more, to our security. Don't stop here! Learn more about electronics. It is important for you, your job, and your future.

Index

A

AC 22
AC circuit 24
AC generator 24
AC power 230
AC/DC adapter 100, 230, 232
AC/DC converter 231
air core 78, 79
alkaline cell 48
alkaline-manganese cell 48
alternating current 22
AM 169
American Wire Gauge 42
ampere 12, 14, 45
amplification factor 112
amplifier, audio
    118, 194, 195, 196, 249, 256
amplifier, operational 229
analog multimeter 221, 223, 224
analog signal 187
anode gate 173
antimatter bodies 6

arsenium 104
atom 3
atomic particles 3
audio amplifier
    118, 194, 195, 196, 249, 256
audio frequency inductor 76
audio oscillator 139
AWG 42

B

ballast 253
bandpass filter 216
barium titanate 92
base 112
base-emitter junction 112
battery 47, 230
battery charger 250
battery test 226
bayonet base 59
Bipolar transistor 111
bipolar transistor
    115, 128, 140, 154, 229

boron 96
Brown and Sharpe (B&S) Wire Gauge 42

C

cadmium sulfide 63
cadmium sulfide photocell 244
capacitance 70, 71, 110
capacitor 37, 69, 100, 160, 167, 184, 229, 230, 241
capacitor, ceramic 69, 73
capacitor, Electrolytic 72
capacitor, electrolytic 18, 74, 229
capacitor, mica 69
capacitor, plastic film 70
capacitor, polyester 257
capacitor, polyester film 73
capacitor, Polystyrene film 73
capacitor, tantalum 72
capacitor, variable 74, 76
carbon electrode 47
carbon-zinc cell 48
cathode 106
CdS cell 63, 65
cell 47
cell, alkaline 48
ceramic capacitor 69, 73
Ceramic transducer 93
choke 76
circuits, electronic 1
closed circuit 9
closed loop 9
CMOS 208, 264
coil 18, 42, 76, 84, 88, 227
collector 112, 124

Collector Current 130
Collector-Emitter Voltage 130
Common Mode Rejection Ratio 193
compression wave 25
conductor 95
continuity test 226
cool solder 34
cosmic ray 29
coupling, Darlington 119
coupling, direct 119
coupling, LC 119
coupling, RC 119
coupling, transformer 119
Critical Rate of Rise 158
crystalline plate 93
current 14

D

Darlington coupling 119
Darlington transistor 128, 132
databook 113
DC 21
DC circuit 24
DC generator 22
DC motor 87
DC power 230
DC voltage 100
decibel 226
decompression wave 25
devices, electronic 2
diac 171, 172
dielectric 69
dielectric material 69
digital multimeter 223, 224
Digital Signal Processor 215

Index

diode 95, 97, 100, 184, 230, 241
diode bridge 161
diode, direct-biased 227
diode test 226
diode, zener 101, 103, 173, 175
direct coupling 119
direct current 21
direct current generator 21
Dissipation Power 130
DPDT 46, 82
DSP 216
dynamo 9

E

earphone 42
electric circuit 8, 9, 14
electric current 5, 7
electric field 26
electric resistance 13
electric tension 12, 14
electric voltage 12
electricity 1
electrode 47, 69
electrolysis 18
electrolyte 47
electrolytic capacitor 18, 72, 74, 229
electromagnet 18, 81, 249
electromagnetic field 26
electromagnetic interference 167
electromagnetic switch 81
electromagnetic wave 17, 24, 26
electromotive force 12
electron 4, 23
electron flow 5, 12

electronic circuit 1, 24
electronic devices 2
electronics 1
Electrostatic Discharge 220
emf 12
EMI 167, 233, 244, 254, 264
emitter 112, 124
enameled wire 42
ENIAC 204
ESD 220

F

farad 70
ferrite 76
ferrite core 79
ferrite toroid 78
FET 140, 148, 229
Field-Effect Transistor 140
filament 58, 60
filament, tungsten 58
filter 230
filter choke 76
flange 59
fluorescent 17
fluorescent lamp 253
FM 169
frequency 26
fuse 43, 227
fusible link 44

G

GaAs 104
gain 130, 193
gallium 95, 104
Gallium Arsenide 104

gamma ray 28
gas, inert 83
gate 140, 145, 152, 153
Gate Trigger Current 165
gate voltage 140
gate-cathode junction 162
gauge 42
generator 8, 9
germanium 95, 97
germanium device 97
germanium transistor 120

H

heat sink 16, 114, 234
henry 76
hertz 24

I

IC 184, 185, 229
IGBT 154
impedance 80, 90, 91
impedance match 80
incandescent lamp 58
indium 96, 104
inductance coil 78
inductor 76
inert gas 83
infrared 17, 29, 107, 123, 247, 260
insulator 95
integrated circuit 179, 184, 187, 196, 216
inverter 234, 235, 245
ionization 17
iron core 78, 79

Isolated-Gate Bipolar Transistor 154

J

JFET 141
joule 15
Joule's Effect 15, 16
Joule's Law 15
Junction FET 141
junction test 120

L

lamp, incandescent 58
lamp, neon 60
laser 247
LC coupling 119
LCD 175, 178, 227
LCD driver 177
LDR 63, 64, 124, 244
LED 17, 60, 104, 105, 108, 123, 241, 252
Light Emitting Diodes 17, 104
light-dependent resistor 63
light-sensitive resistor 63
liquid crystal display 175
loudspeaker 42

M

magnetic field 18, 26, 86
magnetic sensor 42
magnetic transducer 91
match transformer 94
Maxwell, James Clerk 26

Index

megahertz 24
Metal Oxide Varistor 67
Metal-Oxide Semiconductor Field-Effect Transistor 145
mica capacitor 69
microcontroller 213, 214
microfarad 70
microhenry 76
microprocessor 127, 213, 214
miliampere 45
millihenry 76
monochromatic light source 104
MOSFET 145, 146
Motion detector 248
motor 18, 42, 86
multimeter 45, 55, 65, 84, 101, 120, 162, 71, 219, 224, 227
multimeter, analog 221, 223, 224
multimeter, digital 223, 224

N

nanofarad 70
negative charge 7
negative pole 9
Negative Temperature Coefficient 65
neon 17
neon gas 60
neon lamp 60
neutron 3
ni-cad (nickel-cadmium) 48
nonferrous wire 42
NPN 111, 115, 128
NPN silicon transistor 120
NPN transistor 117

NTC 65, 67
nucleus 3

O

Off State Voltage 158
ohm 13, 14, 37, 50
ohmic resistance 91
Ohm's Law 13
operational amplifier 190, 193, 229
optocoupler 123, 127
orbital shell 3
oscillating electric charge 28
oscillator 94, 109, 136, 161, 234, 240, 242
oscillator, audio 139
output transistor 237

P

parallel 20
PCB 30, 46
Phase Locked Loop 199
phosphorus 96
photo sensor 247
photocell 244
photodiode 109, 111, 124, 126, 260
Photofact, Sams 220
phototransistor 122, 124, 126, 247, 260
picofarad 70
piezoelectric 248
piezoelectric ceramic 93
piezoelectric components 94
piezoelectric material 92

piezoelectric transducer 92
piezoelectric unit 89
piroelectric sensor 248
plastic film capacitor 70
PLL 199, 257
PN junction 104, 109, 110
PNP 111, 115, 128
PNP silicon transistor 120
PNP transistor 119
polarity 224, 232, 252
polyester capacitor 257
polyester film capacitor 73
Polystyrene film capacitor 73
positive charge 7
positive pole 9
Positive Temperature Coefficient 65
positron 6
potentiometer 56, 57, 160
potentiometer, slide 57
potentiometer, trimmer 56
power conditioner 255
Power FET 148
power MOSFET 148
preamplifier 118
primary winding 78, 80
printed circuit board 30, 46
probe connections 224
probes 224, 226
proton 3, 4
PTC 65, 67
PUT 139

Q

quadrac 172
quartz crystal 92, 93

R

Radio Frequency Interference 167
radio wave 29
RC coupling 119
rectifier 99, 161, 230
Red-Blue-Green 106
reed switch 83
relay 18, 42, 128, 152
repetitive peak reverse voltage 165
repetitive pulse reverse voltage 165
resistance 13, 14, 219
resistor 13, 37, 50, 229
resistor, fixed 50
resistor, Light-dependent 63
resistor, light-sensitive 63
resistor, variable 50, 56, 57, 74
Resistor, Voltage Dependent 67
reverse-biased diode 227
RF 260
RF inductor 76
RF transformer 80
RFI 167
RGB 106

S

Sams Photofact 220
Sams Technical Publishing 220
schematic 220
SCR 154, 157, 175, 216, 229, 233, 244, 264
secondary winding 78, 80

Index

semiconductor 95, 109, 154, 229
sensor 18, 244, 246, 260
sensor, piroelectric 248
sensor, vibration 249
sensors, photo 247
sequencer 241
series 20
series-parallel associations 21
SHF 109
silicon 95
Silicon Bilateral Switch 174
silicon chip 185, 216
Silicon Controlled Rectifier 154
silicon device 97
Silicon Unilateral Switch 172
sine 24
sine curve 29
slide potentiometer 57
SMD 31
solder 32
soldering gun 34
soldering iron 32
solenoid 18, 42, 85,
    101, 128, 150
solid-state electronics 96
SPDT 47, 82
SPST switch 44
static 5
static electricity 5, 6
step motor 86, 88, 152
Superhigh Frequency 109
surface-mounted devices 31
surge 264
surge protector 264
switch 45, 81
switch, electromagnetic 81

switch, reed 83
switch-mode power supplies
    153, 188

T

tantalum capacitor 72
telephone slide 59
thyristor 137, 154, 161
timbre 26
transducer 18, 88, 91, 245
transducer, Ceramic 93
transducer, magnetic 91
transducer, piezoelectric 92
transformer 78, 80, 230, 241
transformer coupling 119
Transient Surge Absorber 67
Transistor 95
transistor 37, 104, 111,
    124, 216, 234
transistor, Bipolar 111
transistor, bipolar effect 140
Transistor Logic IC 207
transistor manual 113
transistor, output 237
transistor test 226
transistor, unijunction 132
Transition Frequency 130
triac 161, 163, 165, 167,
    172, 216, 229, 233, 244, 264
trigger voltage 171
trimmer potentiometer 56
TTL 207
tube 181
tungsten filament 58
tunnel diode 109, 111
tweeter 90, 93

U

UHF 109, 259, 262
UJT 132
Ultrahigh Frequency 109
ultrasonic sound 25
ultrasound 249
ultrasound transducer 261
ultraviolet 17, 123
unijunction transistor 132
universal power adapter 232

V

varactor 109
variable capacitance diode 110
variable capacitor 74, 76
variable resistor 50, 56, 57, 74
Varicap 109, 110
Varistor, Metal Oxide 67
Varistor, Zinc-Oxide 67
VDR 67, 69, 264
very low frequency 263
VHF 169, 259, 262
vibration sensor 249
visible light 29
VLF 263
volt 12, 14
Volt-Ohm-Milliamp meter 221
voltage 14, 219
Voltage Dependent Resistor 67
voltage multiplier 241
voltage regulator 255
voltage spike 264
voltage-dependent resistor 264
VOM 55, 221

W

watt 15, 90
waveform 26, 37
Winchester disk 91
wire 42
wire, enameled 42
wire, nonferrous 42
woofer 90
Working Voltage DC 71
WVDC 71

X

X-ray 28
xenon 17, 239

Z

zener diode 101, 103, 173, 175
Zinc-Oxide Varistor 67

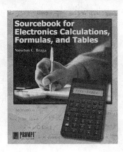

Sourcebook for Electronics Calculations, Formulas, and Tables
Newton C. Braga

Fun Projects for the Experimenter
Newton C. Braga

For the student, teacher, or experienced technician, this is your one-stop guide for formulas and calculations on nearly any electronics subject, including explanations of use, derivations, and practical applcation examples. Topics include cable resistance, electrolysis, basic capacitor forumlas, harmonics, and many more.

Author Newton C. Braga, whose works have appeared in electronics magazines for over 20 years, has collected fifty of his most fun, easy-to-build, and practical projects for your enjoyment. Basic electronic principles and fundamentals are stressed. These projects are primarily stand-alone, low-cost, and with few components, intended for one evening of work. The components needed are listed along with schematics, and hints and questions about the circuits are included to stimulate your imagination regarding possible modifications and alternate use.

Examples of the projects include an LED flasher, mini-metronome, electronic fishing lure, micro FM transmitter, touch switch, wireless beeper, and signal tracer. For hobbyists and students wanting to understand the basic principles of electronics, this book will provide answers to many of your questions.

Professional Reference
440 pages • paperback • 8-3/8 x 10-7/8"
ISBN: 0-7906-1193-7 • Sams 61193
$34.95

Projects
328 pages • paperback • 7-3/8 x 9-1/4"
ISBN: 0-7906-1149-X • Sams 61149
$24.95

To order today or locate your nearest Prompt® Publications distributor at 1-800-428-7267 or www.samswebsite.com

Prices subject to change.

HWS Servicing Series: Zenith Televisions
by Bob Rose

Expanding on the HWS Servicing Series, author Bob Rose takes an in-depth look at Zenith TVs, with coverage of manufacturer history, test equipment, literature, software, and parts. A variety of chassis are given a thorough analysis.

Servicing RCA/GE Televisions
Bob Rose

Let's assume you're a competent technician, with a good work ethic, knowledgeable, having the proper tools to do your job efficiently. You're hungry to learn, to find a way to get an edge on the competition—better yet, to increase your productivity, and hence, your compensation.

In *Servicing RCA/GE Televisions*, author Bob Rose has compiled years of personal experience to share his knowledge about the unique CTC chassis. From the early CTC130 through the CTC195/197 series, Bob reveals the most common faults and quickest ways to find them, as well as some not-so-common problems, quirks and oddities he's experienced along the way. From the RCA component numbering system to the infamous "tuner wrap" problem, Bob gives you all you need to make faster diagnoses and efficient repairs, with fewer call-backs—and that's money in the bank!

Troubleshooting & Repair
352 pages • paperback • 8-1/2" x 11"
ISBN 0-7906-1216-X • Sams 61216
$34.95

Troubleshooting & Repair
352 pages • paperback • 8-3/8" x 10-7/8"
ISBN 0-7906-1171-6 • Sams 61171
$34.95

To order today or locate your nearest Prompt® Publications distributor at 1-800-428-7267 or www.samswebsite.com

Prices subject to change.

Exploring Solid-State Amplifiers
Joseph Carr

Modern Electronics Soldering Techniques
Andrew Singmin

Exploring Solid-State Amplifiers is a complete and authoritative guide to the world of amplifiers. If you're a professional technician or a hobbyist interested in learning more about amplifiers, this is the book for you.

Beginning with amplifier electronics: overcoming the effects of noise, this book covers many useful and interesting topics. It includes helpful, detailed schematics and diagrams to guide you through the circuitry and construction of solid-state amplifiers, such as: Transistor Amplifiers; Junction Field-Effect Transistors and MOSFET Transistors; Operational Amplifiers; Audio Small Signal and Power Amplifiers; Solid-State Parametric Amplifiers; and Monolithic Microwave Integrated Circuits (MMICs). Two bonus chapters are devoted to troubleshooting circuits and selecting solid-state replacement parts.

The traditional notion of soldering no longer applies in the quickly changing world of technology. Having the skills to solder electronics devices helps to advance your career.

Modern Electronics Soldering Techniques is designed as a total learning package, providing an extensive electronics foundation that enhances your electronics capabilities. This book covers how to solder wires and components as well as how to read schematics. Also learn how to apply your newly learned knowledge by following step-by-step instructions to take simple circuits and convert them into prototype breadboard designs. Other tospic covered include troubleshooting, basic math principles used in electronics, simple test meters and instruments, surface-mount technology, safety, and much more!

Electronics Technology
240 pages • paperback • 7-3/8" x 9-1/4"
ISBN: 0-7906-1192-9 • Sams: 61192
$29.95

Electronics Basics
304 pages • paperback • 6" x 9"
ISBN: 0-7906-1199-6 • Sams 61199
$24.95

To order today or locate your nearest Prompt® Publications distributor at 1-800-428-7267 or www.samswebsite.com

Prices subject to change.

SMD Electronics Projects
Homer Davidson

Guide to HDTV Systems
Conrad Persson

SMD components have opened up a brand-new area of electronics project construction. These tiny components are now available and listed in many of the electronics mail-order catalogs for the electronics hobbyist. Projects include: earphone radio, shortwave receiver, baby monitor, cable checker, touch alarm and many more. Thirty projects in all!

As HDTV is developed, refined, and becomes more available to the masses, technicians will be required to service them. Until now, precious little information has been available on the subject. This book provides a detailed background on what HDTV is, the technical standards involved, how HDTV signals are generated and transmitted, and a generalized description of the circuitry an HDTV set consists of. Some of the topics include the ATSC digital TV standard, receiver characteristics, NTSC/HDTV compatibility, scanning methods, test equipment, servicing considerations, and more.

Projects
320 pages • paperback • 7-3/8" x 9-1/4"
ISBN 0-7906-1211-9 • Sams 61211
$29.95

Video Technology
256 pages • paperback • 7-3/8" x 9-1/4"
ISBN 0-7906-1166-X • Sams 61166
$34.95

To order today or locate your nearest Prompt® Publications distributor at 1-800-428-7267 or www.samswebsite.com

Prices subject to change.

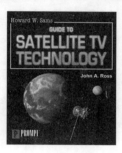

Howard W. Sams Guide to Satellite TV Technology
John A. Ross

This book covers all aspects of satellite television technology in a style that breaks "tech-talk" down into easily understood reading. It is intended to assist consumers with the installation, maintenance, and repair of their satellite systems. It also contains sufficient technical content to appeal to technicians as a reference.

Coverage includes C, Ku, and DBS signals. Chapters include How Satellite Television Technology Works, Parts of a Satellite Television Reception System, Installing the Hardware Portion of Your System, Installing the Electronics of Your System, Installing Your DSS, DBS, or Primestar System, Setting Up a Multi-Receiver Installation, Maintaining the System, Repairing Your System, and more.

Communications
464 pages • paperback • 7-3/8" x 9-1/4"
ISBN 0-7906-1176-7 • Sams 61176
$39.95

Complete Guide to Digital Cameras
Michael Murie

The *Complete Guide to Digital Cameras* will appeal to anyone who has recently purchased or is considering an investment in a digital camera. The first section introduces the reader to digital cameras, how they work, uses and features, and how to buy one. The second section gives tips on use, available options, and how to transfer images from camera to computer. The third section focuses on manipulating the images on computer in varying file formats, and looks at some color printers presently available. Along with model comparisons and index of currently available cameras, a CD-ROM contains sample images, trial software, and utilities.

The *Complete Guide to Digital Cameras* is the answer to all your questions. Author Michael Murie, avid photographer and multimedia developer, has compiled a comprehensive volume, covering all aspects of the digital camera world. As a bonus, we include a CD-ROM with sample images and software, as well as a comprehensive table of the latest digital cameras, comparing features and technical specifications.

Video Technology
536 pages • paperback • 7-3/8" x 9-1/4"
ISBN 0-7906-1175-9 • Sams 61175
$39.95

To order today or locate your nearest Prompt® Publications distributor at 1-800-428-7267 or www.samswebsite.com

Prices subject to change.

Home Automation Basics
Practical Applications Using Visual Basic 6
by Tom Leonik, P.E.

This book explores the world of Visual Basic 6 programming with respect to real-world interfacing, animation and control on a beginner to intermediate level.

Home Automation Basics demonstrates how to interface to a home automation system via the serial port on your PC. Using a programmable logic controller (PLC) as a home monitor, this book walks you through the process of developing the home monitor program using Visual Basic programming. After programming is complete the PLC will monitor the following digital inputs: front and rear doorbell pushbuttons, front and rear door open sensors, HVAC systems, water pumps, mail box, as well as temperature controls. The lessons learned in this book will be invaluable for future serial and animations projects!

Home Automation Basics II: LiteTouch System
James Van Laarhoven

Daunted by the thought of installing a world-class home automation system? You shouldn't be! LiteTouch Systems has won numerous awards for its home automation systems, which combine flexibility with unlimited options. James Van Laarhoven takes LiteTouch to the next level with this text from Prompt® Publications. Van Laarhoven helps to make the installation, troubleshooting, and maintenance of LiteTouch 2000® a less daunting task for the installer, presenting information, examples, and situations in an easy-to-read format. LiteTouch 2000® is a true computer-control system that reduces unsightly switch banks and bulky high-voltage control wiring to a minimum.

Van Laarhoven's efforts make LiteTouch 2000® a product to be utilized by programmers of varying backgrounds and experience levels. *Home Automation Basics II* should be a part of every electrical and security professional's reference library.

Electronics Technology
386 pages • paperback • 7-3/8" x 9-1/4"
ISBN 0-7906-1214-3 • Sams 61214
$34.95

Electronics Technology
304 pages • paperback • 7-3/8" x 9-1/4"
ISBN 0-7906-1226-7 • Sams 61226
$34.95

To order today or locate your nearest Prompt® Publications distributor at 1-800-428-7267 or www.samswebsite.com

Prices subject to change.

Audio Systems Technology, Level I
Handbook for Installers and Engineers
NSCA

This book is a one-stop information source for today's audio technician. It can be used as a study guide to prepare for NICET audio technician certification exams, as well as a comprehensive reference for the installer of audio systems—both out in the field and at the bench.

Designed to correspond with Level I work elements on the NICET tests, this valuable handbook presents the basics of audio installation as it is practiced in the industry today. Topics include: Basic Electronic Circuits; Basic Math; Basics of Microphones & Loudspeakers; Basic Wiring; Switches and Connectors; Codes, Standards and Safety; and Reading Plans and Specifications.

Additionally, information about getting certified by NICET is included, with tips and strategies to help with your success on the NICET exams.

Audio Systems Technology, Level II
Handbook for Installers and Engineers
NSCA

Designed to correspond with Level II work elements on the NICET tests, this book presents intermediate level content on audio installation as it is practiced in the industry today. Some of the topics: Audio Calculations; Acoustical Measurements; Microphones, Loudspeakers and Mixers; Trigonometry and Geometry; Wiring and Cabling; Effective Business Communication; and Bench Test Equipment.

Information about getting certified by NICET is included, with tips and strategies for the test-taker.

Both Level I and Level II books are for you if:
- You want to be NICET-certified in audio systems.
- You are a developing audio installer or engineer.
- You want to enhance your knowledge of audio systems design and installation.
- You are experienced but want to brush up on intermediate-level audio.

Audio
320 pages • paperback • 7-3/8 x 9-1/4"
ISBN: 0-7906-1162-7 • Sams 61162
$34.95

Audio
432 pages • paperback • 7-3/8 x 9-1/4"
ISBN: 0-7906-1163-5 • Sams 61163
$39.95

To order today or locate your nearest Prompt® Publications distributor at 1-800-428-7267 or www.samswebsite.com

Prices subject to change.

Audio Systems Technology, Level III
Handbook for Installers and Engineers
NSCA

    Audio Systems Technology Level III is an essential for the library of the advanced technician who has several years of job experience and an associate's degree or the equivalent. While each book in this series contains its own unique information, there is overlap from one level to the next, providing repetition on the most important, fundamental points. This intentional dovetailing also allows the entire series to be used as a systematic, progressive course of study from the basics through intermediate and advanced topics.

Theory and Design of Loudspeaker Enclosures
J.E. Benson

    Written for design engineers and technicians, students, and intermediate-to-advanced level acoustics enthusiasts, *Theory & Design of Loudspeaker Enclosures* presents a general theory of loudspeaker enclosure systems. Full of illustrated and numerical examples, this book examines diverse developments in enclosure design, and studies the various types of enclosures as well as varying parameter values and performance optimization.

    Topics examined in *Theory & Design of Loudspeaker Enclosures* include: The Synthesis of Vented Systems, Infinite-Baffle and Closed-Box Systems, Electro-Acoustical Relations, Reflex Response Relationships, System Response Formulae, Input Impedance, Circuit Parameters, System Parameters, Driver Parameters, Analogous Circuits, Terminology, And More.

Audio
320 pages • paperback • 7-3/8 x 9-1/4"
ISBN: 0-7906-1178-3 • Sams 61178
$34.95

Audio
244 pages • paperback • 6 x 9"
ISBN: 0-7906-1093-0 • Sams: 61093
$24.95

To order today or locate your nearest Prompt® Publications distributor at 1-800-428-7267 or www.samswebsite.com

Prices subject to change.

Computer Networking for Small Businesses
by John Ross

Small businesses, home offices, and satellite offices have flourished in recent years. These small and unique networks of two or more PCs can be a challenge for any technician. Small network systems are vastly different from their large-office counterparts. Connecting to multiple and off-site offices provides a unique set of challenges that are addressed in this book. Topics include installation, troubleshooting and repair, and common network applications relevant to the small-office environment.

Available in December

Communication
368 pages • paperback • 7-3/8" x 9-1/4"
ISBN 0-7906-1221-6 • Sams: 61221
$39.95

Guide to Webcams
by John Breeden & Jason Byre

Webcams are one of the hottest technologies on the market today. Their applications are endless and cross all demographics. Digital video cameras have become an increasingly popular method of communicating with others across the Internet. From video e-mail clips to web sites that broadcast people's lives for all to see, webcams have become a way for people to battle against the perceived threat of depersonalization caused by computers and to monitor areas from anywhere in the world. Projects and topics include picking a camera, avoiding obstacles, setup and installation, and applications.

Video Technology
320 pages • paperback • 7-3/8" x 9-1/4"
ISBN 0-7906-1220-8 • Sams: 61220
$29.95

To order today or locate your nearest Prompt® Publications distributor at 1-800-428-7267 or www.samswebsite.com

Prices subject to change.

The Complete RF Technician's Handbook, *Second Edition*
Cotter W. Sayre

This is THE handbook for the RF or wireless communications beginner, student, experienced technician or ham radio operator, furnishing valuable information on the fundamental and advanced concepts of RF wireless communications.

Circuits found in the majority of modern RF devices are covered, along with current test equipment and their applications. Troubleshooting down to the component level is heavily stressed. RF voltage and power amplifiers, classes of operation, configurations, biasing, coupling techniques, frequency limitations, noise and distortion are all included, as well as LC, crystal and RC oscillators. RF modulation and detection methods are explained in detail – AM, FM, PM and SSB – and associated circuits, such as AGC, squelch, selective calling, frequency multipliers, speech processing, mixers, power supplies, AFC, multiplex-

Guide to PIC Microcontrollers
by Carl Bergquist

PICs, or peripheral interface controllers, easy to use and are the "chip of the '90s." Aimed at students and entry level technicians, Bergquist displays his expertise in the electronics field with this excellent guide on the use of PIC Microcontrollers. Projects and topics include:
- Electrical Structure
- Software Codes
- Prototype layout boards

Communications Technology
366 pages • paperback • 8-1/2 x 11"
ISBN: 0-7906-1147-3 • Sams: 61147
$34.95

Electronics Technology
336 pages • paperback • 7-3/8" x 9-1/4"
ISBN 0-7906-1217-8 • Sams 61217
$39.95

To order today or locate your nearest Prompt® Publications distributor at 1-800-428-7267 or www.samswebsite.com

Prices subject to change.

Computer Networking for Small Businesses
by John Ross

Small businesses, home offices, and satellite offices have flourished in recent years. These small and unique networks of two or more PCs can be a challenge for any technician. Topics include:
- Installation
- Troubleshooting and repair
- Common network applications

Telecommunications Technologies:
Voice, Data & Fiber Optic Applications
by John Ross

This book contains the information needed to develop a complete understanding of the technologies used within telephony, data and telecommunications networks. Projects and topics include:
- Equipment comparisons
- Business office applications
- Understanding the Technology

Communication
368 pages • paperback • 7-3/8" x 9-1/4"
ISBN 0-7906-1221-6 • Sams 61221
$39.95

Communications
368 pages • paperback • 7-3/8" x 9-1/4"
ISBN 0-7906-1225-9 • Sams 61225
$39.95

To order today or locate your nearest Prompt® Publications distributor at 1-800-428-7267 or www.samswebsite.com

Prices subject to change.

PROMPT
PUBLICATIONS

IC Design Projects
Stephen Kamichik

Build Your Own Test Equipment
Carl J. Bergquist

IC Design Projects discusses some of the most popular and practical ICs, and offers you some projects in which you can learn to create useful and interesting devices with these ICs.

Once you have read through this book and completed some of its projects, you will have a real, working knowledge of ICs, enabling you to design and build you own projects!

Topics include: how power supplies operate, integrated circuit voltage regulators, TTL logic, CMOS logic, how operational amplifiers work, how phase-locked loops work, and more!

Projects include: battery charger, bipolar power supply, capacitance meter, stereo preamplifier, function generator, DC motor control, automatic light timer, darkroom timer, LM567 tone decoder IC, electronic organ, and more!

Test equipment is among the most important tools that can be used by electronics hobbyists and professionals. Building your own test equipment can carry you a long way toward understanding electronics in general, as well as allowing you to customize the equipment to your actual needs.

Build Your Own Test Equipment contains information on how to build several pragmatic testing devices. Each and every device is designed to be highly practical and space conscious, using commonly-available components.

Projects include: Prototype Lab, Multi-Output Power Supply, Signal Generator and Tester, Logic Probe, Transistor Tester, IC Tester, Portable Digital Capacitance Meter, Four-Digit Counter, Digital Multimeter, Digital Function Generator, Eight-Digit Frequency Counter, Solid-State Oscilloscope, and more.

Projects
261 pages • paperback • 7-3/8 x 9-1/4"
ISBN: 0-7906-1135-X • Sams 61135
$24.95

Professional Reference
267 pages • paperback • 7-3/8 x 9-1/4"
ISBN: 0-7906-1130-9 • Sams: 61130
$29.95

To order today or locate your nearest Prompt® Publications distributor at 1-800-428-7267 or www.samswebsite.com

Prices subject to change.

Exploring the World of SCSI
by Louis Columbus

Focusing on the needs of the hobbyist, PC enthusiast, as well as system administrator, *The World of SCSI* is a comprehensive book for anyone interested in learning the hands-on aspects of SCSI. It includes how to work with the Logical Unit Numbers (LUNs) within SCSI, how termination works, bus mastering, caching, and how the various levels of RAID provide varying levels of performance and reliability. This book provides the functionality that intermediate and advanced system users need for configuring SCSI on their systems, while at the same time providing the experienced professional with the necessary diagrams, descriptions, information sources, and guidance on how to implement SCSI-based solutions. Chapters include: How SCSI Works; Connecting with SCSI Devices; and many more.

Dictionary of Modern Electronics Technology
Andrew Singmin

New technology overpowers the old everyday. One minute you're working with the quickest and most sophisticated electronic equipment, and the next minute you're working with a museum piece. The words that support your equipment change just as fast.

If you're looking for a dictionary that thoroughly defines the ever-changing and advancing world of electronic terminology, look no further than the Modern Dictionary of Electronics Technology. With up-to-date definitions and explanations, this dictionary sheds insightful light on words and terms used at the forefront of today's integrated circuit industry and surrounding electronic sectors.

Whether you're a device engineer, a specialist working in the semiconductor industry, or simply an electronics enthusiast, this dictionary is a necessary guide for your electronic endeavors.

Communication
500 pages • paperback • 7-3/8" x 9-1/4"
ISBN 0-7906-1210-0 • Sams 61210
$34.95

Electronics Technology
220 pages • paperback • 7 3/8 x 9 1/4"
ISBN: 0-7906-1164-4 • Sams 61164
$34.95

To order today or locate your nearest Prompt® Publications distributor at 1-800-428-7267 or www.samswebsite.com

Prices subject to change.

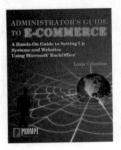

Electronic Projects for the 21st Century
John Iovine

Administrator's Guide to E-Commerce,
A Hands-On Guide to Setting Up Systems and Websites
by Louis Columbus

If you are an electronics hobbyist with an interest in science, or are fascinated by the technologies of the future, you'll find *Electronic Projects for the 21st Century* a welcome addition to your electronics library. It's filled with nearly two dozen fun and useful electronics projects designed to let you use and experiment with the latest innovations in science and technology. This book contains the expert, hands-on guidance and detailed instructions you need to perform experiments that involve genetics, lasers, holography, Kirlian photography, alternative energy sources and more. You will obtain all the information necessary to create the following: biofeedback/lie detector device, ELF monitor, Geiger counter, MHD generator, expansion cloud chamber, air pollution monitor, laser power supply for holography, pinhole camera, synthetic fuel from coal, and much more.

Unlike previous electronic commerce books which stress theory, the *Administrator's Guide to E-Commerce* is a hands-on guide to creating and managing Web sites using the Microsoft BackOffice product suite. A complete guide to setting up your own E-Commerce site.

Projects
256 pages • paperback • 7-3/8 x 9-1/4"
ISBN: 0-7906-1103-1 • Sams: 61103
$24.95

Business Technology
416 pages • paperback • 7-3/8" x 9-1/4"
ISBN 0-7906-1187-2 • Sams 61187
$34.95

To order today or locate your nearest Prompt® Publications distributor at 1-800-428-7267 or www.samswebsite.com

Prices subject to change.

The Component Identifier and Source Book, *2nd Edition*
Victor Meeldijk

Because interface designs are often reverse engineered using component data or block diagrams that list only part numbers, technicians are often forced to search for replacement parts armed only with manufacturer logos and part numbers.

This book was written to assist technicians and designers in identifying components from prefixes and logos, as well as find sources for various types of microcircuits and other components. It will help readers cross reference component types to manufacturers, and also cross reference trade names, abbreviations, and part number prefixes to the manufacturer. Listings of the worldwide manufacturing and sales office addresses are provided, with a special listing of manufacturers who provide replacement devices and vendors who specialize in stocking and locating discontinued devices. There is not another book on the market that lists as many manufacturers of such diverse electronic components.

Professional Reference
425 pages • paperback • 8-1/2 x 11"
ISBN: 0-7906-1159-7 • Sams: 61159
$34.95

Internet Guide to the Electronics Industry
John J. Adams

Although the Internet pervades our lives, it would not have been possible without the growth of electronics. It is very fitting then that technical subjects, data sheets, parts houses, and of course manufacturers, are developing new and innovative ways to ride along the Information Superhighway. Whether it's programs that calculate Ohm's Law or a schematic of a satellite system, electronics hobbyists and technicians can find a wealth of knowledge and information on the Internet.

In fact, soon electronics hobbyists and professionals will be able to access on-line catalogs from manufacturers and distributors all over the world, and then order parts, schematics, and other merchandise without leaving home. The *Internet Guide to the Electronics Industry* serves mainly as a directory to the resources available to electronics professionals and hobbyists.

Professional Reference
192 pages • paperback • 5-1/2 x 8-1/2"
ISBN: 0-7906-1092-2 • Sams: 61092
$19.95

To order today or locate your nearest Prompt® Publications distributor at 1-800-428-7267 or www.samswebsite.com

Prices subject to change.

Desktop Digital Video
by Ron Grebler

The Video Hacker's Handbook
by Carl Bergquist

Desktop Digital Video is for those people who have a good understanding of personal computers and want to learn how video (and digital video) fits into the bigger picture. This book will introduce you to the essentials of video engineering, and to the intricacies and intimacies of digital technology. It examines the hardware involved, then explores the variety of different software applications and how to utilize them practically. Best of all, *Desktop Digital Video* will guide you through the development of your own customized digital video system. Topics covered include the video signal, digital video theory, digital video editing programs, hardware, digital video software and much more.

Geared toward electronic hobbyists and technicians interested in experimenting with the multiple facets of video technology, *The Video Hacker's Handbook* features projects never seen before in book form. Video theory and project information is presented in a practical and easy-to-understand fashion, allowing you to not only learn how video technology came to be so important in today's world, but also how to incorporate this knowledge into projects of your own design. In addition to the hands-on construction projects, the text covers existing video devices useful in this area of technology plus a little history surrounding television and video relay systems.

Video Technology
225 pages • paperback • 7-3/8 x 9-1/4"
ISBN: 0-7906-1095-7 • Sams: 61095
$34.95

Video Technology
336 pages • paperback • 7-3/8 x 9-1/4"
ISBN: 0-7906-1126-0 • Sams: 61126
$29.95

To order today or locate your nearest Prompt® Publications distributor at 1-800-428-7267 or www.samswebsite.com

Prices subject to change.